I0030546

NASA

NATIONAL AERONAUTICS AND SPACE ADMINISTRATION
WASHINGTON, D.C. 20546

TELS. WO 2-4155
WO 3-6925

RELEASE NO: 71-3K

FOR RELEASE: THURSDAY A.M.
January 21, 1971

P
R
E
S
S

K
I
T

PROJECT: APOLLO 14

(To be launched no
earlier than Jan. 31)

contents

-more-

1/11/71

ILLUSTRATIONS

Cover: The Glare of the Sun
This view of Antares, the Apollo 14 Lunar Module as it sat on the moon's Fra Mauro Highlands, reflects a circular flare caused by the brilliant sun. The unusual ball of light was said by the astronauts to have a jewel-like appearance. At extreme left, the lower slope of Cone Crater can be seen.
Image Credit: NASA

Published by Books Express Publishing
Copyright © Books Express, 2012
ISBN 978-1-78039-863-1

Books Express publications are available from all good retail and online booksellers. For publishing proposals and direct ordering please contact us at: info@books-express.com

NEWS   **NASA**   NATIONAL AERONAUTICS AND SPACE ADMINISTRATION   (202) 962-4155
WASHINGTON, D.C. 20546   **TELS:**   (202) 963-6925

FOR RELEASE: THURSDAY A.M.
January 21, 1971

RELEASE NO: 71-3

APOLLO 14 LAUNCH JAN. 31

Apollo 14, the sixth United States manned flight to the Moon and fourth Apollo mission with an objective of landing men on the Moon, is scheduled for launch Jan. 31 at 3:23 p.m. EST from Kennedy Space Center, Fla.

The Apollo 14 lunar module is to land in the hilly upland region north of the Fra Mauro crater for a stay of about 33 hours, during which the landing crew will leave the spacecraft twice to set up scientific experiments on the lunar surface and to continue geological explorations. The two earlier Apollo lunar landings were Apollo 11 at Tranquillity Base and Apollo 12 at Surveyor 3 crater in the Ocean of Storms.

Apollo 14 prime crewmen are Spacecraft Commander Alan B. Shepard, Jr., Command Module Pilot Stuart A. Roosa, and Lunar Module Pilot Edgar D. Mitchell. Shepard is a Navy captain Roosa an Air Force major and Mitchell a Navy commander.

-more-   1/8/71

Lunar materials brought back from the Fra Mauro formation are expected to yield information on the early history of the Moon, the Earth and the solar system--perhaps as long ago as five billion years.

During their two Moonwalks, Shepard and Mitchell will set up a series of experiments, the Apollo Lunar Surface Experiments Package (ALSEP) and will conduct extensive geological surveys of the area around the landing site. The crew will be aided by a two-wheeled pull-cart carrying experiments and geology tools during their lengthy field geology trek.

Experiments in the ALSEP are: Passive Seismic for long-term measurement of lunar seismic events; Active Seismic for relaying to Earth data on the lunar crust; the Suprathermal Ion Detector and Cold Cathode Ion Gauge for measuring ion flux, density and energy in the lunar environment; and a Charged Particle Lunar Environment Experiment for measuring energy of solar protons and electrons reaching the Moon and a Portable Magnetometer for measuring variations in the lunar magnetic field in the geology traverse will be carried on the lunar cart. The crew will set up a laser beam reflector, similar to the one left by the Apollo 11 crew, for long-term observatory measurements of Earth-Moon distance and motion relationships.

While the commander and lunar module pilot are exploring the Fra Mauro area, the command module pilot will be carrying out several orbital science tasks in lunar orbit above, including photography of dim-light phenomena and candidate landing sites. The candidate landing sites will be photographed with a large-format lunar topographic camera mounted in the hatch window which takes high-resolution photos for overlapping stereo sequences or for photomosaics.

Also, photos of earlier Apollo landing sites will be aboard for correlation with previous tracking data to improve tracking accuracy techniques.

The Apollo 14 flight profile in general follows those flown by Apollos 11 and 12 with two major exceptions: Lunar orbit insertion burn No. 2 has been combined with descent orbit insertion and the docked spacecraft will be placed into a 10 by 58-nautical mile lunar orbit by the service propulsion system.

Lunar module propellant is conserved by combining these maneuvers and by using the service module engine to provide 15 seconds of additional hover time during the landing.

Also, additional tracking time in the descent orbit provides more accurate position and velocity data for use in the landing. The other change is in the lunar orbit rendezvous. Many of the intermediate maneuvers leading up to rendezvous and docking after LM ascent stage liftoff have been omitted, and rendezvous will take place shortly before the end of the first revolution after ascent.

Lunar surface touchdown is scheduled for 4:16 a.m. EST Feb. 5, and two periods of lunar surface extravehicular activity are planned at 8:53 a.m. EST on the 5th and 5:38 a.m. EST Feb. 6. The LM ascent stage will lift off at 1:47 p.m. EST on the 6th to rejoin the lunar orbiting command module after about 33 hours on the lunar surface.

Apollo 14 will leave lunar orbit for the Earthward trip at 8:37 p.m. EST Feb. 6. Splashdown in the South central Pacific, south of American Samoa, will be at 4:01 p.m. EST Feb. 9.

After the command module has landed, the crew will don clean coveralls and filter masks passed to them through the hatch by a swimmer, and then transfer by helicopter to a Mobile Quarantine Facility (MQF) aboard the USS New Orleans. Later the crew will fly by helicopter to Samoa, then board another MQF on a C-141 aircraft for the flight back to the Lunar Receiving Laboratory at the Manned Spacecraft Center in Houston.

-more-

The crew will remain in quarantine up to 21 days from completion of the second EVA.

The Apollo 14 crew selected the callsigns "Kitty Hawk" for the command/service module and "Antares" for the lunar module. When all three men are aboard the command module, the callsign will be "Apollo 14." As in the two previous lunar landing missions, an American flag will be emplaced on the lunar surface. A plaque bearing the date of the Apollo 14 landing and the crew signatures is attached to the LM.

Apollo 14 backup crewmen are USN Capt. Eugene A. Cernan, commander; USN Comdr. Ronald E. Evans, command module pilot; and USAF Lt. Col. Joe H. Engle, lunar module pilot.

-end-

COUNTDOWN

A government-industry team of about 500 will conduct the Apollo 14 countdown from Firing Room 2 of the Launch Control Center, at Kennedy Space Center's Complex 39.

Precount activities begin at T-6 days when final preparations start for the space vehicle official countdown. During this period, space vehicle pyrotechnics and electrical connections are completed. Mechanical buildup of spacecraft components is accomplished, along with servicing the various gases and cryogenics (liquid oxygen and liquid hydrogen) to the CSM and the LM for the mission. The spacecraft batteries are installed and the fuel cells are activated.

The official countdown begins at T-28 hours and continues to T-9 hours, at which time a built-in hold is planned prior to the start of launch vehicle propellant loading.

Following are some of the highlights of the later parts of the countdown:

T-10 hours, 15 minutes	Start Mobile Service Structure (MSS) move to park site
T-9 hours	Built-in hold for nine hours and 23 minutes. At end of hold, pad is cleared for LV propellant loading
T-8 hours, 05 minutes	Launch vehicle propellant loading - Three stages (LOX in first stage, LOX and LH_2 in second and third stages). Continues thru T-3 hours 38 minutes
T-4 hours, 17 minutes	Flight crew alerted
T-4 hours, 02 minutes	Crew medical examination
T-3 hours, 32 minutes	Brunch for crew
T-3 hours, 30 minutes	One-hour built-in hold

T-3 hours, 07 minutes	Crew departs Manned Spacecraft Operations Building for LC-39 via transfer van
T-2 hours, 55 minutes	Crew arrival at LC-39
T-2 hours, 40 minutes	Start flight crew ingress
T-1 hour, 51 minutes	Space Vehicle Emergency Detection System (EDS) test (Shepard participates along with launch team)
T-43 minutes	Retract Apollo access arm to stand-by position (12 degrees)
T-42 minutes	Arm launch escape system
T-40 minutes	Final launch vehicle range safety checks (to 35 minutes)
T-30 minutes	Launch vehicle power transfer test, LM switch over to internal power
T-20 minutes to T-10 minutes	Shutdown LM operational instrumentation
T-15 minutes	Spacecraft to full internal power
T-6 minutes	Space vehicle final status checks
T-5 minutes, 30 seconds	Arm destruct system
T-5 minutes	Apollo access arm fully retracted
T-3 minutes, 6 seconds	Firing command (automatic sequence)
T-50 seconds	Launch vehicle transfer to internal power
T-8.9 seconds	Ignition sequence start
T-2 seconds	All engines running
T-0	Liftoff

Note: Some changes in the countdown are possible as a result of experience gained in the countdown demonstration test which occurs about two weeks before launch.

LAUNCH AND MISSION PROFILE

The Apollo 14 space vehicle is scheduled for liftoff from Launch Complex 39A at the Kennedy Space Center, Fla., at 3:23 p.m. EST, January 31, 1970, on an azimuth of 72 degrees.

The Apollo 14 trajectory in general will follow the same profile flown in previous lunar landing missions. Initially, the Apollo 14 spacecraft will be on a trajectory that would allow the spacecraft to coast past the Moon and return to Earth if no further major maneuvers were made.

At about 30.5 hours ground elapsed time, (GET) the spacecraft will maneuver to a trajectory to produce the desired conditions of altitude, time, and sun-angle at lunar orbit insertion. This new trajectory is constrained to permit an Earth return with the lunar module descent engine should the SPS engine not restart.

At about 82.5 hours GET a retrograde SPS burn will place Apollo 14 into a 57x170 nm lunar orbit, followed by a descent orbit insertion (DOI) SPS burn at about 87 hours GET which will place the entire spacecraft into a 10x58 nm lunar orbit.

On Apollos 11 and 12, DOI was a separate maneuver using the LM descent engine while the CSM remained in a 60-nm circular orbit. A total spacecraft DOI maneuver to the low pericynthion orbit was planned for the Apollo 13 mission.

The new DOI maneuver in effect is a combination lunar orbit insertion burn No. 2 (LOI-2) and DOI, and produces two benefits: conserves LM descent prepellant that would have been used for DOI and makes this propellant available for additional LM hover time near the lunar surface, and allows 11 lunar revolutions of spacecraft tracking in the descent orbit to enhance position/velocity (state vector) data for updating the LM guidance computer during the lunar descent and landing phase.

The LM ascent and rendezvous sequence will differ somewhat from the profiles flown on Apollos 11 and 12 in that the "direct ascent" technique will be used. The goal of the early rendezvous is to insert the LM into a 51x9 nm orbit so that at a predetermined time (38 min.) after orbit insertion the terminal phase initiation (TPI) can be performed using the APS. The total time from insertion to rendezvous will be approximately 85 minutes. The early rendezvous liftoff will occur approximately 2 1/2 minutes prior to the nominal liftoff time for a coelliptic rendezvous. The liftoff window duration is limited to 30 seconds to keep the resultant perilune above 8 nm.

The transearth injection (TEI) maneuver and transearth coast midcourse correction sequence will follow the "standard" pattern. Command module splashdown will be in the South Central Pacific at 216.38 GET at coordinates 27.2° south latitude by 171.5° west longitude.

The Apollo 14 mission events table on the following summarizes the maneuver sequence.

Launch Opportunities

The Apollo 14 Launch windows and lunar landing Sun elevation angles are presented in the following table.

Launch Date	Windows (EST) Open	Close	Sun Elevation Angle
January 31, 1971	3:23 pm	7:12 pm	10.3°
March 1, 1971 (T-24)	3:03 pm	7:07 pm	10.5°
March 2, 1971 (T-0)	3:43 pm	7:38 pm	10.5°
March 3, 1971 (T+24)	4:08 pm	7:47 pm	23.0°
March 30, 1971 (T-24)	2:22 pm	5:57 pm	8.0°
March 31, 1971 (T-0)	2:35 pm	6:00 pm	8.0°
*April 1, 1971 (T+24)	2:45 pm	6:01 pm	22.0°

*Under investigation

Ground Elapsed Time Update

It is planned to update, if necessary, the actual ground elapsed time (GET) during the Apollo 14 mission to allow the major flight plan events to occur at the pre-planned GET regardless of either a late liftoff or trajectory dispersions that would otherwise have changed the event times.

For example, if the flight plan calls for descent orbit insertion (DOI) to occur at GET 86 hours, 57 minutes and the flight time to the Moon is two minutes longer than planned due to trajectory dispersions at translunar injection, the GET clock will be turned back two minutes during the translunar coast period so that DOI occurs at the pre-planned time rather than at 86 hours, 59 minutes. It follows that the other major mission events would then also be accomplished at the pre-planned times.

Updating the GET clock will accomplish in one adjustment that would otherwise require separate time adjustments for each event. By updating the GET clock, the crew and ground flight control personnel will be releived of the burden to change their checklists, flight plans, etc.

The planned times in the mission for updating GET will be kept to a minimum and will, generally, be limited to two updates. If required, they will occur at about 55 hours into the mission and at a time just prior to LM activation. Both the actual GET and the update GET will be maintained in the MCC throughout the mission.

Launch Events

Time Hrs	Min	Sec	Event	Vehicle Wt (Pounds)	Altitude (Feet)	Velocity (Ft/Sec)	Range (Nau Mi)
00	00	00	First Motion	6,423,753	195	1,341	0
00	01	25	Maximum Dynamic Pressure	3,970,879	44,487	2,701	3
00	02	15	S-IC Center Engine Cutoff	2,532,589	137,942	6,192	24
00	02	45	S-IC Outboard Engines Cutoff	1,843,548	219,311	8,938	51
00	02	46	S-IC/S-II Separation	1,836,231	221,546	8,970	52
00	02	47	S-II Ignition	1,468,613	225,856	8,960	54
00	03	15	S-II Aft Interstage Jettison	1,398,378	300,905	9,364	88
00	03	21	Launch Escape Tower Jettison	1,372,573	315,748	9,482	95
00	07	44	S-II Center Engine Cutoff	642,459	592,467	18,575	595
00	09	17	S-II Outboard Engines Cutoff	466,308	614,993	22,861	882
00	09	18	S-II/S-IVB Separation	465,994	615,295	22,873	886
00	09	21	S-IVB Ignition	366,415	616,161	22,874	897
00	11	43	S-IVB First Cutoff	300,580	627,789	25,562	1,416
00	11	53	Parking Orbit Insertion (103 nm)	300,453	627,805	25,568	1,455

Mission Events

Events	GET hrs:min	Date/EST	Velocity change feet/sec	Purpose and resultant orbit
Translunar injection (S-IVB engine ignition)	02:30	31/5:53 pm	10,031	Injection into translunar trajectory with 2030 nm pericynthion
CSM separation, docking	02:51	31/6:14 pm	--	Mating of CSM and LM
Ejection from SLA	03:56	31/7:19 pm	1	Separates CSM-LM from S-IVB/SLA
S-IVB evasive maneuver	04:19	31/7:42 pm	10	Provides separation prior to S-IVB propellant dump and thruster maneuver to cause lunar impact
S-IVB Propellant dump (5.5 min.)	04:42	31/8:05 pm		
S-IVB APS impact burn (4 min.)	06:30	31/9:53 pm		
S-IVB APS corrective burn	09:00	1/12:23 am		
Midcourse correction 1	TLI +9 hrs	1/2:53 am	*0	*These midcourse corrections have a nominal velocity change of 0 fps, but will be calculated in real time to correct TLI dispersions. MCC-2 is an SPS maneuver (73 fps) to lower pericynthion to 57 nm; trajectory within capability of docked DPS burn should SPS fail to ignite.
Midcourse correction 2 (hybrid transfer)	30:36	1/9:59 pm	73	
Midcourse correction 3	LOI -22 hrs	3/4:01 am	*0	
Midcourse correction 4	LOI -5 hrs	3/9:01 pm	*0	
Lunar orbit insertion	82:38	4/2:01 am	-2,986	Inserts Apollo 14 into 57x170nm elliptical lunar orbit.
S-IVB impacts lunar surface	83:06	4/2:29 am	--	Seismic event for Apollo 12 passive seismometer

Events	GET hrs:min	Date/EST	Velocity change feet/sec	Purpose and resultant orbit
Descent orbit insertion (DOI)	86:56	4/6:19 am	-207	SPS burn places CSM/LM into 10x58 nm lunar orbit.
CSM-LM undocking	104:27	4/11:50 pm	--	
CSM circularization	105:46	5/1:09 am	73	Inserts CSM into 56x63 nm orbit (SPS burn.)
LM Powered descent initiation	108:42	5/4:05 am	-6637	Three-phase DPS burn to brake LM out of transfer orbit, vertical descent and touch-down on lunar surface.
LM touchdown on lunar surface	108:53	5/4:16 am	--	Lunar exploration, deploy ALSEP, collect geological samples, photography.
Depressurize for first lunar surface EVA	113:30	5/8:53 am	--	
CDR steps to surface	113:47	5/9:10 am		
CDR unstows Modularized Equipment Transporter (MET)	113:54	5/9:17 am		
LMP steps to surface	114:14	5/9:37 am		
CDR unstows and erects S-Band antenna	114:19	5/9:42 am		
CDR deploys Solar Wind Composition experiment	114:30	5/9:53 am		
LMP reenters LM to switch to surface S-Band antenna	114:32	5/9:55 am		
Crew deploys U.S. Flag	114:42	5/10:05 am		
Crew deploys MET	114:54	5/10:17 am		

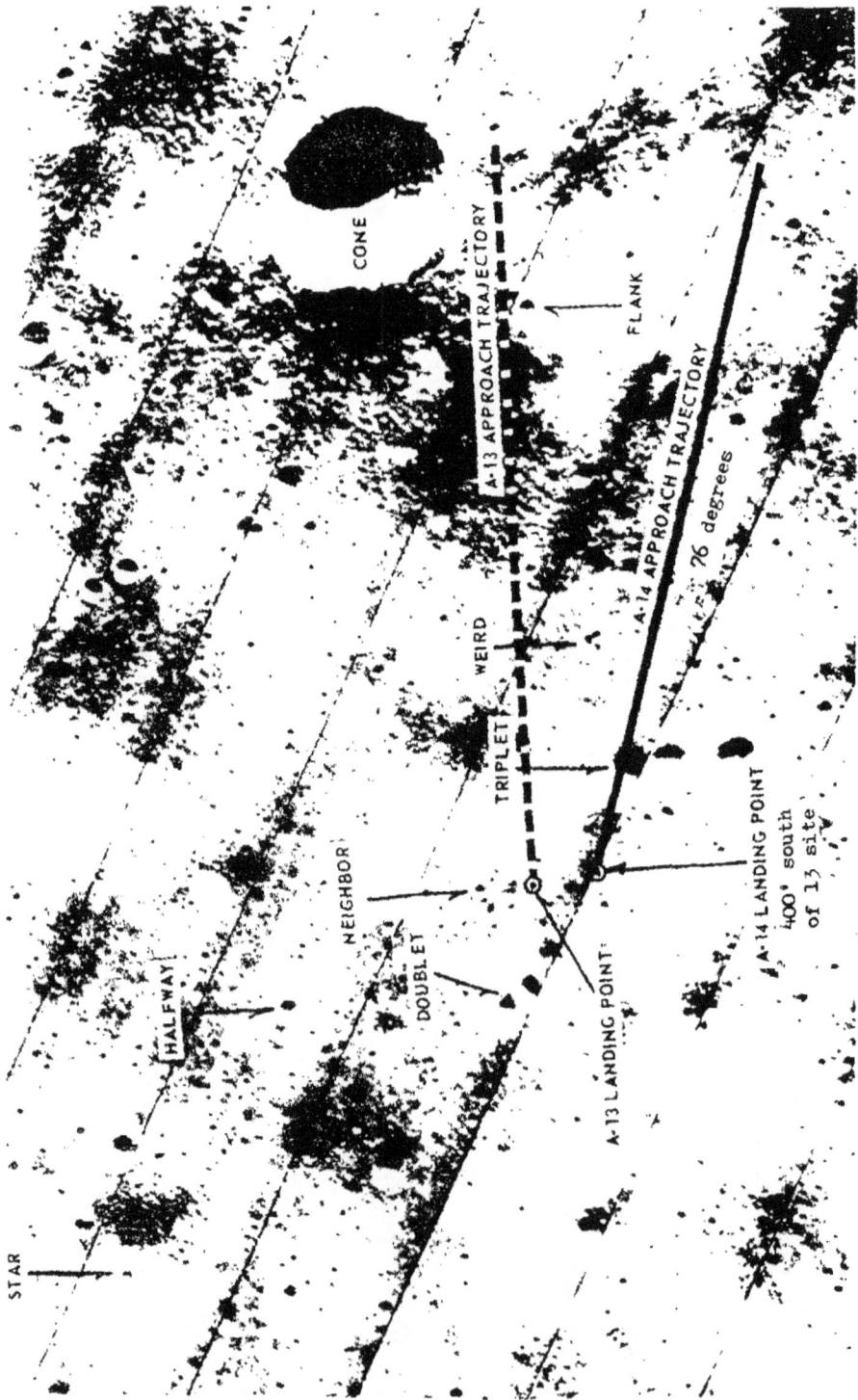

FRA MAURO LANDING SITE

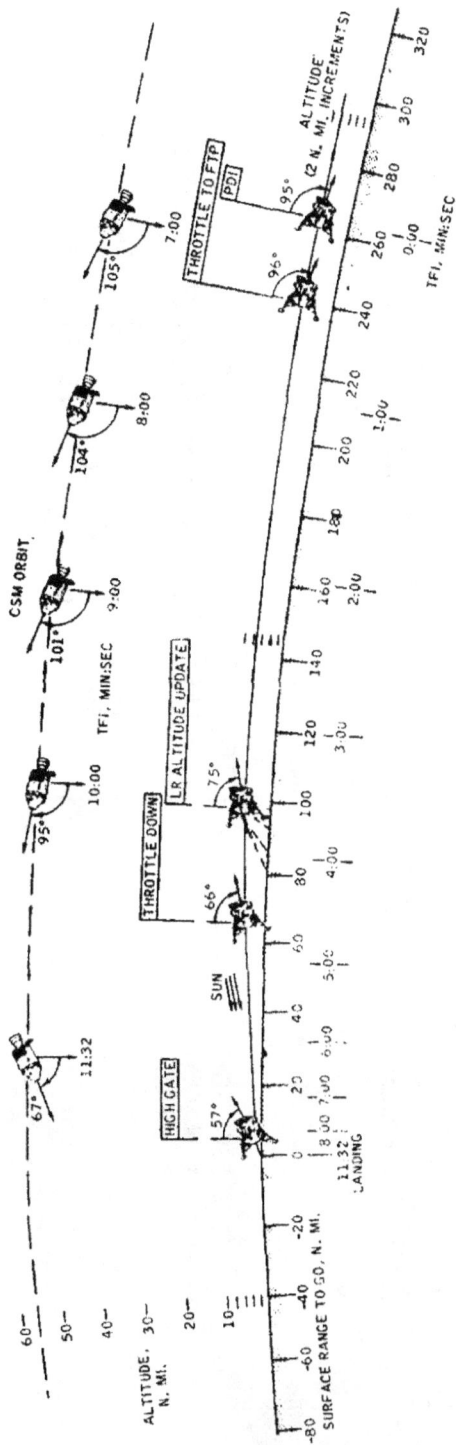

POWERED DESCENT PROFILE

EVENT	TFI, MIN:SEC	V, FPS	Ḣ, FPS	H, FT	ΔV, FPS
POWERED DESCENT INITIATION	0:00	5560	-2	50 023	0
THROTTLE TO MAXIMUM THRUST	0:26	5526	-3	47 964	34
LANDING RADAR ALTITUDE UPDATE	3:36	3601	-89	37 779	1 992
LANDING RADAR VELOCITY UPDATE	5:12	2500	-94	25 763	3164
THROTTLE RECOVERY	6:34	1439	-65	21 430	4289
HIGH GATE	8:34	450	-153	7 692	5436
LOW GATE	10:12	70 (82)*	-20	659	6184
LANDING	11:32	-15 (0)*	-3	5	6637

*(HORIZONTAL VELOCITY RELATIVE TO SURFACE)

SUMMARY

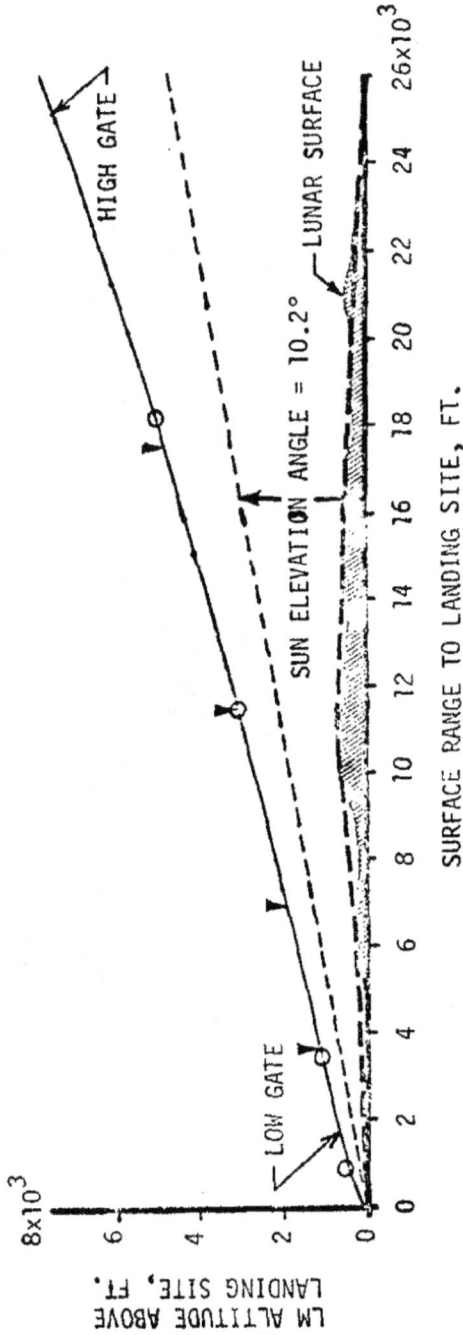

TIME TICKS EVERY 20 SEC. FROM HIGH GATE TO LOW GATE.

HIGH GATE

LUNAR SURFACE

SUN ELEVATION ANGLE = 10.2°

LOW GATE

LM ALTITUDE ABOVE LANDING SITE, FT.

8x10³

SURFACE RANGE TO LANDING SITE, FT.

DIFFERENCES
DESCENT PROFILE
TERRAIN PROFILE
P-66 AUTO

TIME FROM HIGH GATE, SEC	ALTITUDE RATE, FPS	THRUST ANGLE, DEG	HORIZONTAL VELOCITY, FPS
0	-153	57	438
20	-109	27	342
40	-71	28	265
60	-46	28	193
80	-30	26	130
98*	-20	22	82

*LOW GATE

APPROACH PHASE SUMMARY

Cone Crater estimated 330' above landing point and to be 250' deep. External slope 13° (est.)

CONE

EVA TRAVERSE

EVA Distances (Estimated)
EVA-1 by Cdr 3000'
     by LMP 2500'
EVA-2 8920' each

WEIRD

TERMATE

TRIPLET

TRIPLET

SAMPLE AREA

DOUBLET

Events	GET hrs:min	Date/EST	Velocity change feet/sec	Purpose and resultant orbit
Crew offloads ALSEP	115:00	5/10:23 am		
Crew arrives at ALSEP deployment site	115:33	5/10:56 am		
Crew photographs deployed ALSEP and laser reflector	116:30	5/11:53 am		
CDR collects samples in ALSEP area while LMP activates mortar	116:55	5/12:18 pm		
Crew begins return traverse to LM, collects samples en route	117:00	5/12:23 pm		
Crew arrives at LM, stows equipment, samples	117:15	5/12:38 pm		
Crew ingresses LM, repressurizes cabin, end EVA 1	117:45	5/1:08 pm		
CSM plane change No. 1	118:09	5/1:32 pm	36	Changes CSM orbital plane by 3.9° to coincide with LM orbital plane at time of ascent from surface
LM cabin depressurized for EVA 2	134:15	6/5:38 am		
CDR steps to surface	134:28	6/5:51 am		
CDR stows sample gathering equipment on MET	134:35	6/5:58 am		
LMP steps to surface	134:40	6/6:03 am		
CDR evaluates MET, LMP deploys Lunar Portable Magnetometer	134:45	6/6:08 am		
CDR conducts Thermal Degradation Sample experiment	134:55	6/6:18 am		

Events	GET hrs:min	Date/EST	Velocity change feet/sec	Purpose and resultant orbit
Crew collects core tube samples	135:12	6/6:35 am		
Crew begins geology traverse, collecting samples, making stereo and panorama photos en route	135:19	6/6:42 am		
Crew arrives at Cone Crater rim	136:05	6/7:28 am		
Crew arrives back at LM, collects contaminated sample beneath LM, stows samples and film magazines for loading in LM.	137:49	6/9:12 am		
Crew ingresses LM, repressurize LM cabin	138:28	6/9:51 am		
LM ascent	142:24	6/1:47 pm	6053	Boosts ascent stage into 9x51 nm lunar orbit for rendezvous with CSM.
Insertion into lunar orbit	142:31	6/1:54 pm		
Terminal phase initiate (TPI) (LM APS)	143:09	6/2:32 pm	78	Boosts ascent stage into 61x44 nm catch-up orbit; LM trails CSM by 29 nm and 15 nm below at time of TPI burn.
Braking (LM RCS)	143:50	6/3:13 pm	33	Line-of-sight terminal phase braking to place LM in 58x60 nm orbit for final approach, docking
Docking	144:10	6/3:33 pm		CDR and LMP transfer back to CSM

CSM ATTITUDES ARE FOR ILLUSTRATION PURPOSES ONLY

EVENT	TFI, MIN:SEC	INERTIAL VELOCITY, FPS	ALTITUDE RATE, FPS	ALTITUDE, FT	ΔV, FPS	WEIGHT, LB	PROPELLANT, LB
LIFT-OFF	0:00	15.1	0	0	0	10 747.0	0
END OF VERTICAL RISE	0:10	55.7	54.2	269.4	107.5	10 630.8	115.4
BEGIN LM PILOT YAW	0:48	324.8	113.8	3 726.5	526.0	10 190.3	552.6
END LM PILOT YAW	1:00	435.0	126.5	5 169.0	661.6	10 051.5	690.3
	2:00	1036.0	172.5	14 279.2	1366.9	9 360.5	1376.2
	4:00	2474.5	184.8	36 727.0	2934.6	7 990.3	2736.0
	6:00	4272.9	105.9	54 969.6	4779.3	6 632.3	4083.8
ORBIT INSERTION	7:10.7	5540.6	31.3	59 932.7	6053.4	5 831.2	4878.5

SUMMARY

ORBIT INSERTION PHASE

APOLLO 14

EARLY RENDEZVOUS

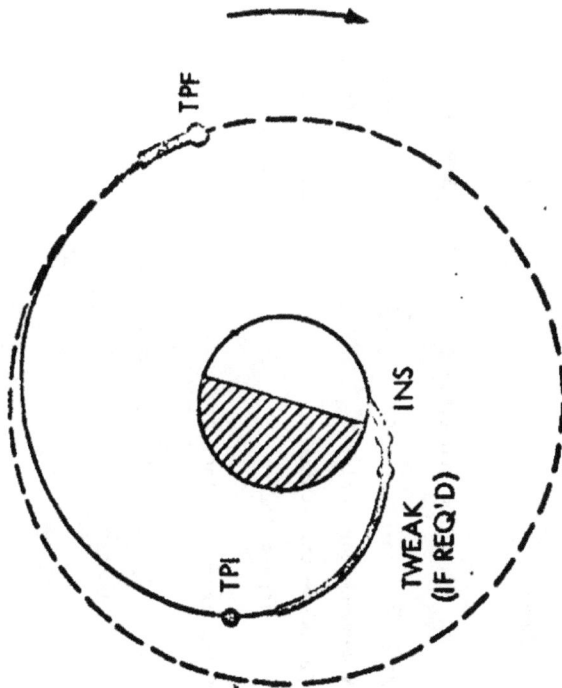

COELLIPTIC RENDEZVOUS

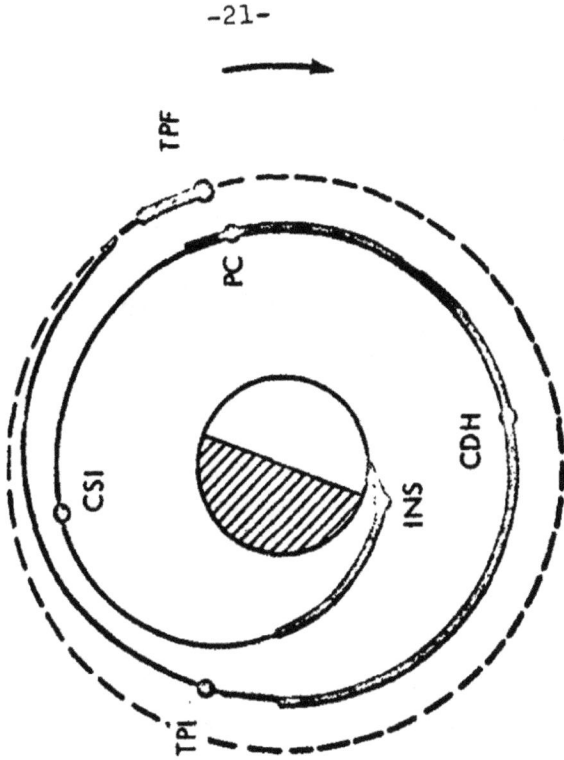

MSFN COVERAGE

RENDEZVOUS PLANS

Events	GET hrs:min	Date/EST	Velocity change feet/sec	Purpose and resultant orbit
LM jettison, separation	146:23	6/5:46 pm		Prevents recontact of CSM with LM ascent stage during remainder of lunar orbit.
LM ascent stage deorbit (RCS)	147:52	6/7:15 pm	-184	ALSEP seismometers at Apollo 14 and Apollo 12 landing sites record impact event.
LM impact	148:20	6/7:43 pm		Impact at about 5506 fps at -4° angle, 29 nm from Apollo 14 ALSEP
Transearth injection (TEI) SPS	149:14	6/8:37 pm	3450	Inject CSM into transearth trajectory
Midcourse correction 5	TEI +17 hrs	7/1:37 pm	0	Transearth midcourse corrections will be computed in real time for entry corridor control and recovery area weather avoidance.
Midcourse correction 6	EI -22 hrs	8/5:49 pm	0	
Midcourse correction 7	EI -3 hrs	9/12:49 pm	0	
CM/SM separation	216:11	9/3:34 pm		Command module oriented for Earth atmosphere entry.
Entry interface (400,000 ft)	216:24	9/3:47 pm		Command module enters atmosphere at 36,170 fps
Splashdown	216:38	9/4:01 pm		Landing 1250 nm downrange from entry.splash at 27.2°Slat by 171.5°WLong.

Entry Events

Event	Time from 400,000 ft, min:sec	
Entry	00:00	3:47 p.m. 9th
Enter S-band communication blackout	00:18	
Initiate constant drag	00:52	
Maximum heating rate	01:10	
Maximum load factor (FIRST)	01:20	
Exit S-band communication blackout	03:32	
Maximum load factor (SECOND)	06:16	
Termination of CMC guidance	07:16	
Drogue parachute deployment	08:16	
Main parachute deployment	09:05	
Landing	13:54	4:01 p.m. 9th

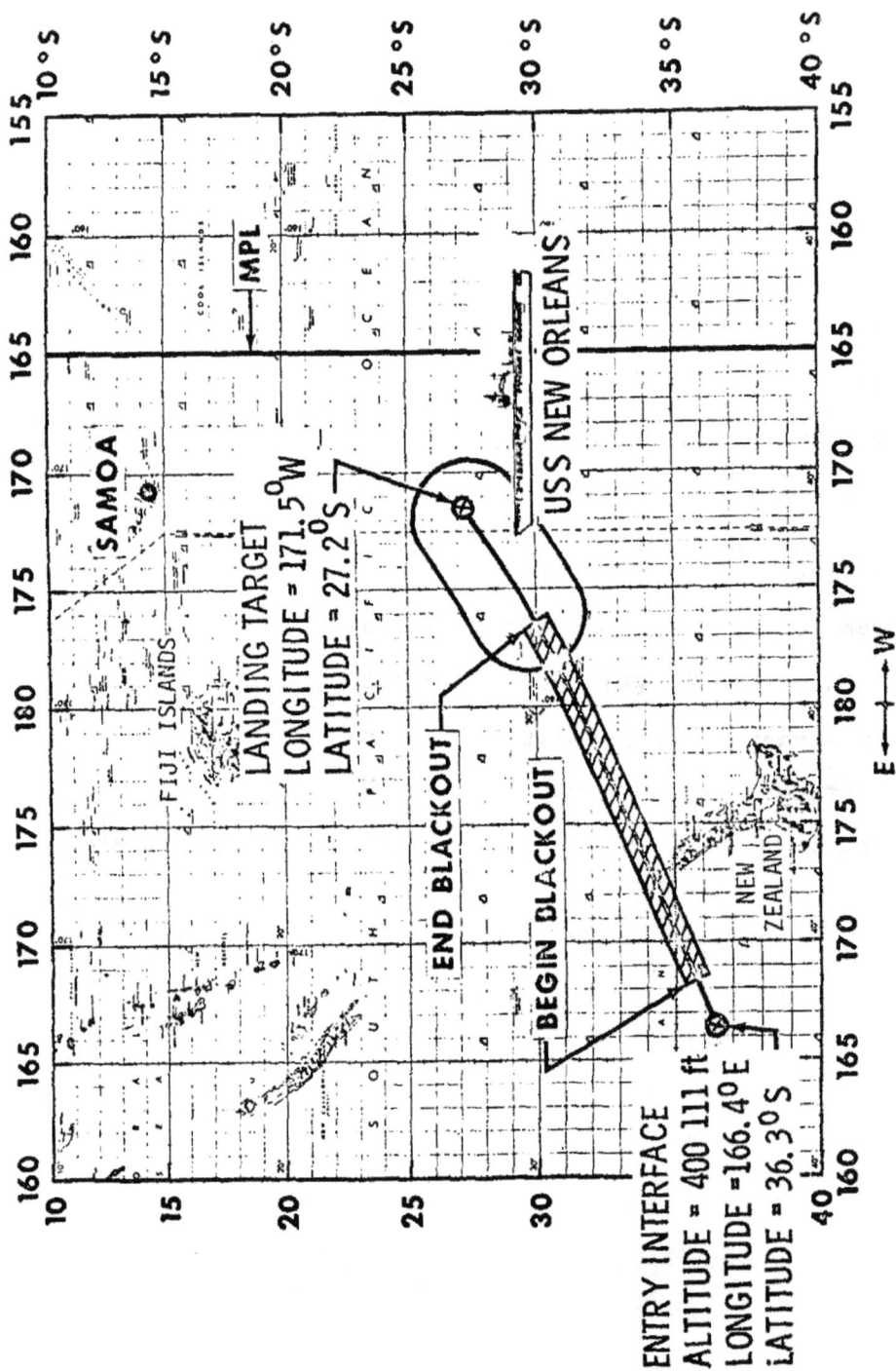

PRIMARY LANDING AREA

RECOVERY OPERATIONS

Launch abort landing areas extend downrange 3,400 nautical miles (nm) from Kennedy Space Center, fanwise 50 nm above and below the limits of the variable launch azimuth (72-96 degrees) in the Atlantic Ocean. On station in the launch abort area will be the destroyer USS Hawkins.

Splashdown for a full-duration lunar landing mission launched on time Jan. 31 will be at 27.2° South latitude by 171.5° West longitude, about 780 nm south of Pago Pago, American Samoa, and about an equal distance southeast of Suva, Fiji Islands.

The landing platform-helicopter (LPH) USS New Orleans, Apollo 14 prime recovery vessel, will be stationed near the Pacific Ocean end-of-mission aiming point prior to entry.

In addition to the primary recovery vessel located on the mid-Pacific recovery line and the surface vessel in the launch abort area, seven HC-130 air rescue aircraft will be on standby at five staging bases around the Earth: Guam, Hawaii, Azores, Ascension Island and Florida.

Apollo 14 recovery operations will be directed from the Recovery Operations Control Room in the Mission Control Center, supported by the Atlantic Recovery Control Center, Norfolk, Va. and the Pacific Recovery Control Center, Kunia, Hawaii.

After splashdown, the Apollo 14 crew will don clean coveralls and filter masks passed to them through the spacecraft hatch by a recovery swimmer. The crew will be carried by helicopter to the New Orleans where they will enter a Mobile Quarantine Facility (MQF) about 90 minutes after landing.

Later the crew will travel by helicopter to Samoa, enter another MQF on a C-141 aircraft for flight to the Lunar Receiving Laboratory at Houston. The crew will remain in quarantine up to 21 days from completion of the second EVA.

Crew and Sample Return Schedule

DATE	EVENT
Feb. 9	Splashdown and recovery operations
Feb. 11	Crew and samples depart prime recovery ship MQF for Samoa via helicopter
Feb. 11	First samples arrive Lunar Receiving Laboratory via Samoa and Hawaii
Feb. 11	Crew in MQF and second samples depart Samoa, arrive Hawaii via C-141 aircraft
Feb. 12	Crew in MQF and second samples depart Hawaii, arrive Houston via C-141 aircraft
Feb. 12	Crew in MQF and second samples arrive at LRL
Feb. 12 - 25	Crew debriefings scheduled
Feb. 26	Crew release from quarantine

MISSION OBJECTIVES

The objectives of the Apollo 14 mission are lunar surface
science, lunar orbital science, and operational/engineering.
Lunar surface scientific activities are centered around deployment of the Apollo Lunar Scientific Experiment Package (ALSEP),
lunar field geology investigations, collection of samples of
surface materials for return to Earth, and deployment of other
scientific instruments not part of ALSEP.

The orbital science primarily concerns high-resolution
photography of candidate future landing sites, photography of
deep-space phenomena such as Zodiacal light and Gegenschein;
communications tests using S-band and VHF signals to determine
reflective properties of the lunar surface; tests to determine
variations in lunar gravity at orbital altitude by observing
Doppler variations in S-band signals; and photography of surface
details from 60 nautical miles altitude. The command module
pilot will conduct the bulk of these orbital tasks while the
commander and lunar module pilot are on the lunar surface.

Engineering/operational evaluations of hardware and techniques will continue throughout the mission to add to the
experience fund to be drawn upon in future missions and programs.

Lunar Surface Science

As in previous lunar landing missions, a contingency sample
of lunar surface material will be collected during the first
EVA period.

The Apollo 14 landing crew will devote the major portion
of the first EVA to deploying the ALSEP instruments. These
instruments will remain on the Moon to transmit scientific data
to the Manned Space Flight Network on long-term physical and
environmental properties of the Moon. These data can be correlated
with known Earth data for further knowledge on the origins of
the planet and its satellite.

ALSEP Array C carried on Apollo 14 contains five experiments.
They are: Passive Seismic Experiment (PSE), Active Seismic
Experiment (ASE), Suprathermal Ion Detector (SIDE), Cold Cathode
Ion Gauge (CCIG),and Charged Particle Lunar Environmental
Experiment (CPLEE). Additionally, one independent experiment
will be deployed in the vicinity of the ALSEP, the Laser Ranging
Retro-Reflector (LRRR or LR^3). The Lunar Portable Magnetometer
(LPM) experiment will be conducted during the second EVA.

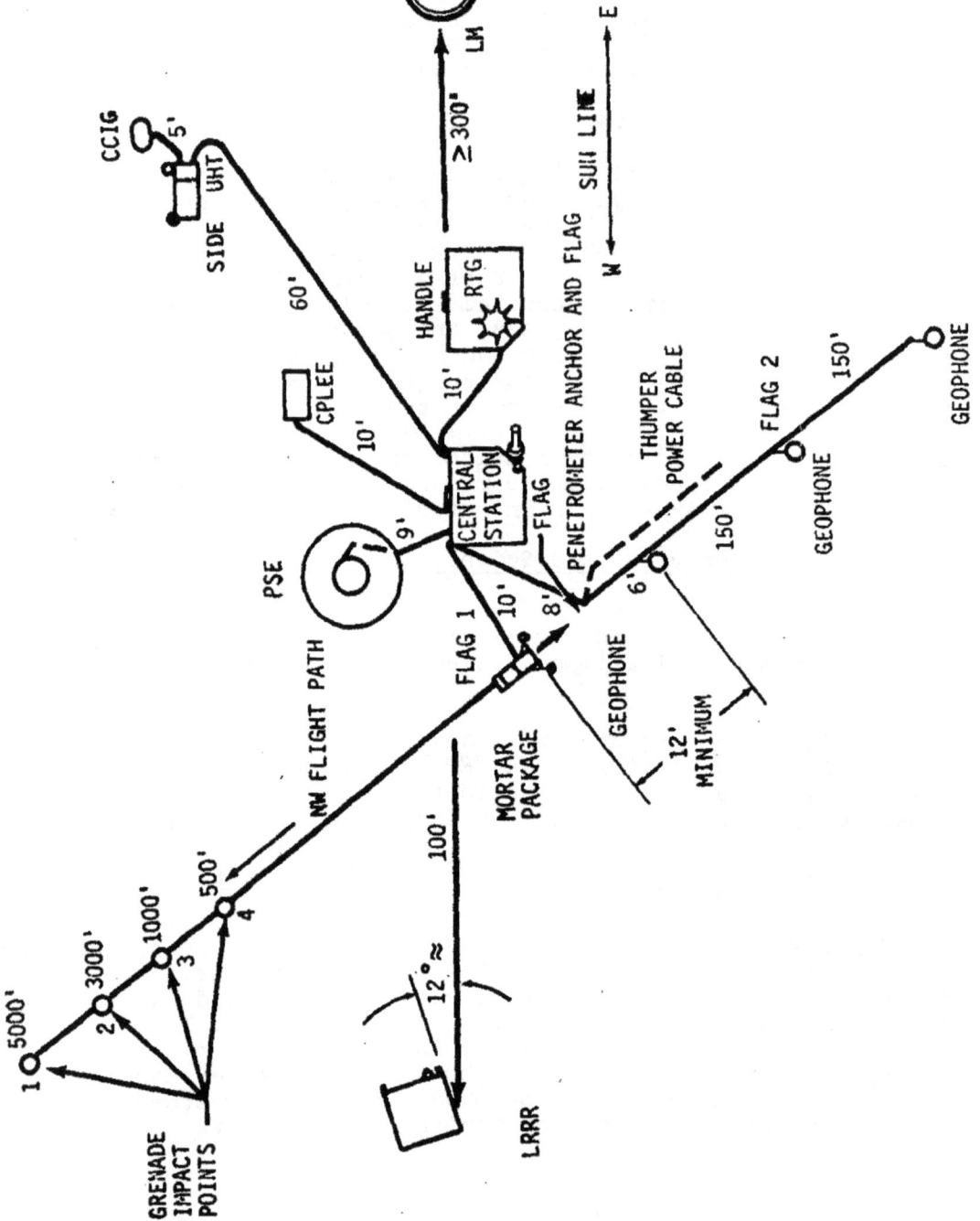

ALSEP ARRAY "C"

Passive Seismic Experiment: The PSE will measure seismic activity of the Moon and gather and relay to Earth information on the physical properties of the lunar crust and interior. The PSE will report seismic data on man-made impacts (LM ascent stage), natural impacts of meteorites, and Moonquakes. Dr. Gary Latham of the Lamont-Doherty Geological Observatory is responsible for PSE design and subsequent experiment analysis.

A similar PSE, deployed as a part of the Apollo 12 ALSEP near Surveyor crater in November 1969, has transmitted to Earth data on lunar surface seismic events since that time. The Apollo 12 and 14 seismometers differ from the seismometer left at Tranquillity Base in July 1969 by the Apollo 11 crew in that the PSEs are continuously powered by a SNAP-27 radioisotope electric generator. The Apollo 11 seismometer, powered by solar cells, could transmit data only during the lunar day, and is no longer functioning.

After Apollo 14 translunar injection, an attempt will be made to impact the spent S-IVB stage and the instrument unit into the Moon. This will stimulate the passive seismometer left on the lunar surface by the Apollo 12 crew in November 1969.

Through a series of switch-selector-commanded and ground-commanded thrust operations, the S-IVB/IU will be directed to hit the Moon within a target area 379 nautical miles in diameter. The target point is between Lansberg FC and Lansberg FB craters (1 degree, 36 minutes south latitude; 33 degrees, 15 minutes west longitude).

After the lunar module is ejected from the S-IVB, the launch vehicle will fire an auxiliary propulsion system (APS) ullage motor to safely separate the vehicle from the spacecraft. Residual liquid oxygen in the almost spent S-IVB/IU will then be dumped through the engine with the vehicle positioned so the dump will slow it into an impact trajectory. Mid-course corrections will be made with the stage's APS ullage motors if necessary.

The S-IVB/IU will weigh 30,836 pounds and will be traveling 4,942 nautical-miles-an-hour at impact. It will provide an energy source at impact equivalent to about 11 tons of TNT.

After Shepard and Mitchell have rendezvoused with the command module in lunar orbit, the lunar module ascent stage will be jettisoned and later ground-commanded to impact on the lunar surface about 32 stature miles from the Apollo 14 landing site at Fra Mauro.

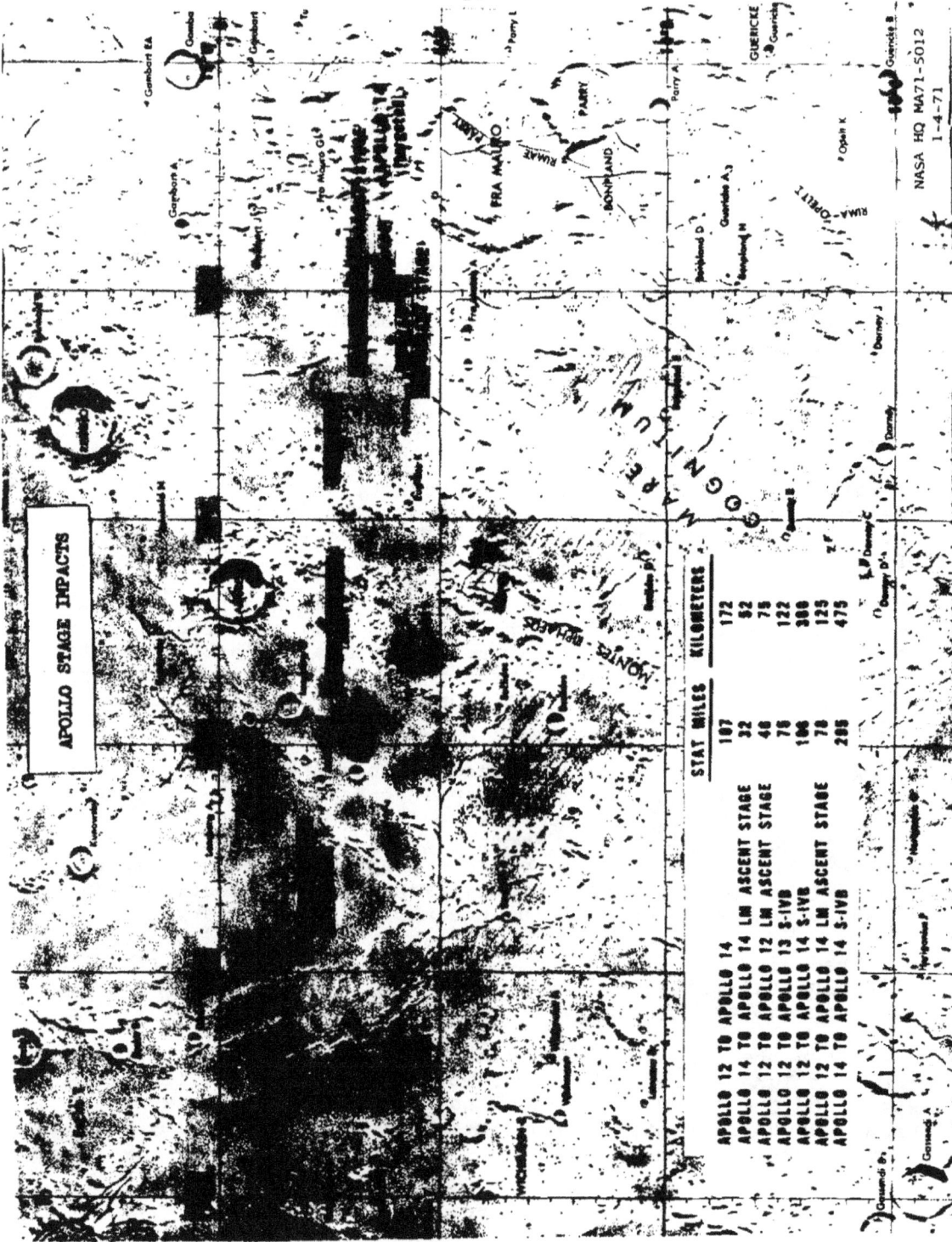

APOLLO STAGE IMPACTS

	STAT MILES	KILOMETERS
APOLLO 12 TO APOLLO 14	107	172
APOLLO 14 TO APOLLO 14 LM ASCENT STAGE	32	52
APOLLO 12 TO APOLLO 12 LM ASCENT STAGE	48	78
APOLLO 12 TO APOLLO 13 S-IVB	75	122
APOLLO 12 TO APOLLO 14 S-IVB	106	300
APOLLO 12 TO APOLLO 14 LM ASCENT STAGE	75	125
APOLLO 14 TO APOLLO 14 S-IVB	295	475

MARE COGNITUM

MONTES RIPHAEUS

FRA MAURO

GUERICKE

NASA HQ MA71-5012
1-4-71

Impact of an object of known mass and velocity will assist in calibrating the Apollo 14 PSE readouts as well as providing comparative readings between the Apollo 12 and 14 seismometers forming the first two stations of a lunar surface seismic network.

There are three major physical components of the PSE:

1. The sensor assembly consists of three long-period and one short-period vertical seismometers with orthagonally-oriented capacitance-type seismic sensors, capable of measuring along two horizontal components and one vertical component. The sensor assembly is mounted on a gimbal platform. A magnet-type sensor short-period seismometer is located on the base of the sensor assembly.

2. The leveling stool allows manual leveling of the sensor assembly by the crewman to within ± 5 degrees. Final leveling to within ± 3 arc seconds is accomplished by control motors.

3. The five-foot diameter hat-shaped thermal shroud covers and helps stabilize the temperature of the sensor assembly. Also, two radioisotope heaters will protect the instrument from the extreme cold of the lunar night.

Active Seismic Experiment: The ASE will produce data on the physical structure and bearing strength of the lunar surface by measuring seismic waves. Two types of man-made seismic sources will be used with the ASE: a crew-actuated pyrotechnic "thumper" and a mortar-like device from which four rocket propelled projectiles can be launched by command from Earth. Naturally produced seismic events will be detected passively by the ASE (the ASE will be turned on remotely for short periods). The seismic waves are detected by geophones deployed by the crew. Data on wave penetration, frequency spectra, and velocity to lunar depths of 500 feet will be obtained and passed to the ALSEP central station for transmittal to the Earth. Dr. Robert Kovach of Stanford University is the Principal Investigator.

The mortar like device will be deployed, aligned and activated about 10 feet northwest of the ALSEP central station. Present plans call for the four grenade-like projectiles to be launched sometime after the crew returns.

The crew will deploy three geophones at 10, 160 and 310 feet from the ALSEP central station. Enroute back to the central station, the crewman will fire 21 "thumper" charges at 15-foot intervals along the geophone line. The thumper serves as a storage and transport rack for the geophones and their connecting cable.

ASE

Projectile

Projectile	-1	-2	-3	-4
RANGE (FT)	5000	3000	1000	500
WEIGHT (GM)	1236	1020	810	719
HIGH EXPLOSIVE CHARGE WEIGHT (LB)	1.0	0.60	0.30	0.10
ROCKET MOTOR MEAN PEAK THRUST (LB)	4600	2575	1575	1200
MEAN VELOCITY (F.P.S.)	161	132	74	53
LUNAR FLIGHT TIME (SEC)	44	32	19	13
ROCKET MOTOR PROPELLANT WEIGHT (GM)	47	31	16.8	11.5
NUMBER OF PELLETS (PROPELLANT)	2435	1596	648	550
LAUNCH ANGLE	45°	45°	45°	45°
ROCKET MOTOR THRUST DURATION (MSEC)	7.0	8.2	12.5	12.5

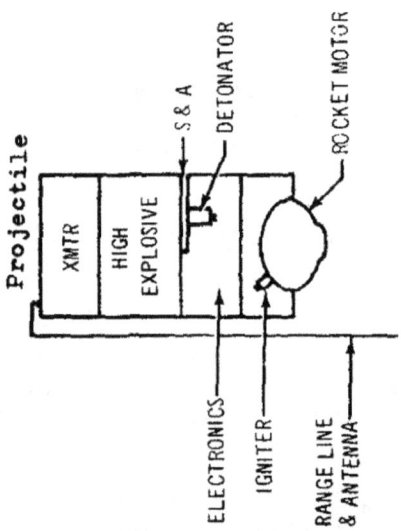

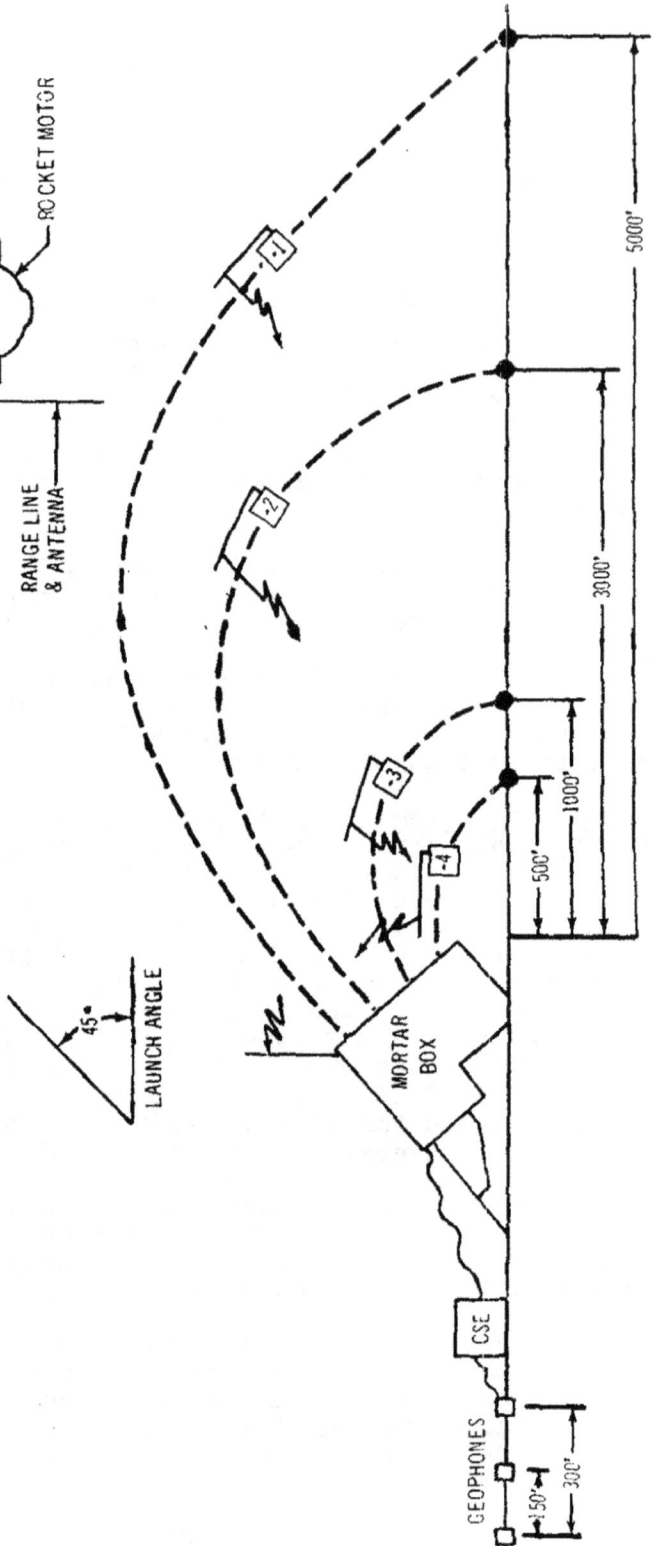

Projectile

XMTR

HIGH EXPLOSIVE

S & A

DETONATOR

ROCKET MOTOR

ELECTRONICS

IGNITER

RANGE LINE & ANTENNA

45° LAUNCH ANGLE

ASE RANGE DATA

5000' 3000' 1000' 500'

-1 -2 -3 -4

MORTAR BOX

CSE

GEOPHONES

300' 150'

-more-

The two major components of the ASE are:

1. The thumper-geophone assembly measuring 44.5 inches when deployed and weighing 7 lbs, including three geophones and cable. Each geophone is 4.8 inches high, 1.6 inches in diameter and weighs less than one pound.

2. The package projectile launch assembly weighs 15 pounds (including 4 projectiles) and is nine and a half inches high, four inches wide and 15.6 inches long.

The mortar-like launching device is made of fiberglass and magnesium, and contains firing circuitry and a receiver antenna. The projectile launch assembly is enclosed in a box and consists of four fiberglass launch tubes and four projectiles. The projectiles vary in length and weight according to the propellant and explosive charges. Radio transmitters in each projectile furnish start-and-stop flight time data for telemetry back to Earth. Thus, with the launch angle known, range can be calculated. The geophones provide information on seismic wave travel time. Correlation of this time with range will establish wave velocity through the lunar surface.

Suprathermal Ion Detector Experiment and Cold Cathode Ion Gauge Experiment: The SIDE will measure flux, composition, energy and velocity of low-energy positive ions and the high-energy solar wind flux of positive ions. Combined with the SIDE is the Cold Cathode Ion Gauge Experiment (CCIG) for measuring the density of the lunar ambient atmosphere and any variations with time or solar activity such atmosphere may have.

Data gathered by the SIDE will yield information on: (1) interaction between ions reaching the Moon from outer space and captured by lunar gravity and those ions that escape; (2) whether or not secondary ions are generated by ions impacting the lunar surface; (3) whether volcanic processes exist on the Moon; (4) effects of the ambient electric field; (5) loss rate of contaminants left in the landing area by the LM and the crew; and (6) ambient lunar atmosphere pressure.

Dr. John Freeman of Rice University is the SIDE principal investigator, and Dr. Francis S. Johnson of the University of Texas is the CCIG principal investigator.

The SIDE instrument consists of a velocity filter, a low-energy curved-plate analyzer ion detector and a high-energy curved-plate analyzer ion detector housed in a case measurng 15.2 by 4.5 by 13 inches, a wire mesh ground plane, and electronic circuitry to transfer data to the ALSEP central station. The SIDE case rests on folding tripod legs. Dust covers, released by ground command, protect both instruments. Total SIDE weight is 19.6 pounds.

The SIDE and the CCIG, connected by a short cable, will be deployed about 55 feet southeast of the ALSEP central station, with the SIDE aligned east or west toward the subearth point and the CCIG orifice aligned along the north-south line pointed toward the Fra Mauro formation with a clear field away from other ALSEP instruments and the LM.

Charged Particle Lunar Environment Experiment: The CPLEE measures particle energies of solar protons and electrons that reach the lunar surface. The instrument will provide data on energy distribution of these solar particles and contribute toward a greater understanding of their effect on the Earth-Moon system; ie. relationship of the solar wind to Earth auroras, Van Allen belt radiation, and other terrestrial phenomena; processes taking place in the shock front of solar wind striking lunar surface; characteristics of the Earth's magnetic tail as it is swept downstream by the solar wind; and effect of charged particles upon the lunar environment. The CPLEE measures protons and electrons in the energy range of 40,000 to 70,000 electron volts (40-70 Kev). Dr. Brian J. O'Brian of the University of Sydney (Australia) is the CPLEE principal investigator.

The insulated CPLEE case measures 10.3 by 4.5 by 10 inches and weighs five pounds. Two spectrometer packages are housed in the case, each oriented for minimum exposure to the Sun's ecliptic path. Each of the two spectrometers has six particle detectors: five C-shaped channeltron photon-multipliers consisting of glass capillary tubes one millimeter inside diameter and 10 centimeters long, and one helical photon multiplier (funneltron) with an eight millimeter opening. Particles of a given charge and different energies entering the spectrometer are subject to varying voltages and deflected toward the five photon multipliers, while particles of the opposite charge are deflected toward the funneltron photon multiplier. Electrons and protons are thus measured simultaneously in five different energy levels. Voltages are shifted in six steps--$\pm$35v, $\pm$350v and $\pm$3500v--to measure electrons and protons from 40 ev to 70 Kev. The CPLEE will be deployed ten feet northeast of the ALSEP central station.

ALSEP Central Station: The ALSEP Central Station serves as a power-distribution and data-handling point for experiments carried on each version of the ALSEP. Central Station components are the data subsystem, helical antenna, experiment electronics, power conditioning unit and dust detector. The Central Station is deployed after other experiment instruments are unstowed from the pallet.

The Central Station data subsystem receives and decodes uplink commands, times and controls experiments, collects and transmits scientific and engineering data downlink, and controls the electrical power subsystem through the power distribution and signal conditioner.

The modified axial helix S-band antenna receives and transmits a right-hand circularly-polarized signal. The antenna is manually aimed with a two-gimbal azimuth/elevation aiming mechanism. A dust detector on the Central Station, composed of three solar cells, measures the accumulation of lunar dust on ALSEP instruments.

The ALSEP electrical power subsystem draws electrical power from a SNAP-27 (Systems for Nuclear Auxiliary Power) radio-isotope thermoelectric generator.

Laser Ranging Retro-Reflector (LRRR) Experiment: The LRRR will permit long-term measurements of the Earth-Moon distance by acting as a passive target for laser beams directed from observatories on Earth. Data gathered from these measurements of the round trip time for a laser beam (accurate to within 15 cm) will be used in the study of fluctuations in the Earth's rotation rate, wobbling motions of the Earth on its axis, the Moon's size and orbital shape, and the possibility of a slow decrease in the gravitational constant "G". Dr. James Faller of Wesleyan University, Middletown, Conn., is LRRR principal investigator.

The LRRR is a square array of 100 fused silica reflector cubes mounted in an adjustable support structure which will be aimed toward Earth by the crew during deployment. Each cube reflects light beams back in absolute parallelism in the same direction from which they came.

By timing the round trip time for a laser pulse to reach the LRRR and return, observatories on Earth can calculate the exact distance from the observatory to the LRRR location within a tolerance of 15 cm. A similar LRRR was deployed at Tranquillity Base by the Apollo 11 crew as an experiment in the Early Apollo Scientific Experiments Package (EASEP). The goal is to set up LRRRs at three lunar locations to establish absolute control points in the study of Moon motion.

Lunar Portable Magnetometer: The LPM will be used by the crew during the second EVA for measuring variations in the lunar magnetic field at several points in the geology traverse. Data gathered will be used to determine the location, strength, and dimensions of the magnetic sources, as well as knowledge of the local selenological structure.

The LPM will be carried on the modularized equipment transporter (MET) and consists of a flux-gate magnetometer sensor head mounted on a tripod and an electronics/data package. The sensor head is connected to the data package by a 50-foot flat cable, and after the lunar module pilot aligns the sensor head at least 35 feet from the data package, he returns to the Mobile Equipment Transporter (MET) and relays readouts to Earth by voice.

The sensor head is sensitive to the magnetic field of the crewman's Portable Life Support System (PLSS), hence the need for aligning the sensor near the end of the cable. The mercury-cell powered electronics package has a high ($\pm$ 100 gamma) and a low ($\pm$ 50 gamma) switchable meter range. Readings are displayed in three taut-band meters--one for each axis (orthoganal X, Y and Z.) At each location for LPM measurements during the traverse, the lunar module pilot will call out the meter readings in each axis at one-minute intervals.

Dimensions of the LPM components are 4 by 7 1/2 by 5 inches for the data package and 3-3/8 by 5-11/16 by 2-5/8 inches for the sensor head. The sensor head tripod is 18 inches long collapsed and extends to 31 inches. The principal investigator is Dr. Palmer Dyal, Ames Research Center, Calif.

Lunar Field Geology: The lunar surface field geology experiment gathers data for interpreting landing site geological history, such as the nature of the origin of the debris layer or regolith, and the land forms superimposed at later dates on the maria and highlands. The lunar bedrock and structure and the types of materials found at the site are expected to yield an insight into the internal processes of the Moon's formation.

One large rock, several smaller fragments and fine-grained material typical of the Fra Mauro site will be collected during EVA 1 to insure the return of samples should the second EVA be cancelled for some reason. The selected samples will be stowed in Sample Return Container No. 1 and taken into the LM at the end of EVA 1.

In addition to the samples gathered by the crew during the two EVA periods and returned to Earth for anlysis, subjective crew comments in real time and photographs (stereo and regular color) and postflight crew debriefing will be the primary means of data gathering.

Specific types of lunar surface samples to be collected during the field geology traverse include six core-tube soil samples, a lunar-environment soil sample from beneath the surface, and a sample of exhaust-contaminated soil from beneath the LM. Dr. Gordon Swann, United States Geological Survey, is principal investigator.

Specific timelines and traverse routes for both EVAs are in the Apollo 14 Lunar Surface Procedures, available for reference at the KSC and MSC News Center query desks.

Soil Mechanics Experiment: The soil mechanics experiment actually is spread out among most of the Apollo 14 activities on the lunar surface, and is aimed toward determining the mechanical characteristics of the lunar surface and subsurface. A trench will be dug by the crew for determinigg the natural angle of repose of excavated material and the sidewall integrity of the excavation itself. Data gained from the trenching task will be useful in lunar shelter and vehicle mobility design.

Additionally, comments and photographs will be made of the interaction of the LM footpads, crew boots, ALSEP instruments and other pieces of gear with the lunar surface. Natural features such as slopes, boulders, ridges, rills, crater walls and embankments will also be observed and photographed. Lunar material characteristics data on texture, consistency, compressibility, cohesion, adhesion, density and color will be gathered.

About three kilograms (6.6 lbs) of fine-grained fragmental material will be brought back for soil mechanics studies in the Lunar Receiving Laboratory. Dr. James Mitchell, University of California, Berkeley, is principal investigator.

Solar Wind Composition Experiment: The scientific objective of the solar wind composition experiment is to determine the elemental and isotopic composition of the noble gases in the solar wind.

As in Apollos 11 and 12, the SWC detector will be deployed on the lunar surface and brought back to Earth by the crew. The detector will be exposed to the solar wind flux for 25 hours compared to 17 hours on Apollo 12 and two hours on Apollo 11.

The solar wind detector consists of an aluminum foil four square feet in area and about 0.5 mils thick rimmed by Teflon for resistance to tearing during deployment. A staff and yard arrangement will be used to deploy the foil and to maintain the foil approximately perpendicular to the solar wind flux. Solar wind particles will penetrate into the foil, allowing cosmic rays to pass through. The particles will be firmly trapped at a depth of several hundred atomic layers. After exposure on the lunar surface, the foil is rolled up and returned to Earth. Professor Johannas Geiss, University of Berne, Switzerland, is principal investigator.

SNAP-27 -- Power Source for ALSEP: A SNAP-27 unit will provide power for the ALSEP package. SNAP-27 is one of a series of radioisotope thermoelectric generators, or atomic batteries developed for the Atomic Energy Commission under its space SNAP program. The SNAP (Systems for Nuclear Auxiliary Power) program is directed at development of generators and reactors for use in space, on land, and in the sea.

While nuclear heaters were used in the seismometer package on Apollo 11, SNAP-27 on Apollo 12 marked the first use of a nuclear electrical power system on the Moon. The use of SNAP-27 on Apollo 14 will mark the second use of such a unit on the Moon.

The basic SNAP-27 unit is designed to produce at least 63.5 electrical watts of power. The SNAP-27 unit is a cylindrical generator, fueled with the radioisotope plutonium-238. It is about 18 inches high and 16 inches in diameter, including the heat radiating fins. The generator, making maximum use of the lightweight material beryllium, weighs about 28 pounds unfueled.

The fuel capsule, made of a superalloy material, is 16.5 inches long and 2.5 inches in diameter. It weighs about 15.5 pounds, of which 8.36 pounds represent fuel. The plutonium-238 fuel is fully oxidized and is chemically and biologically inert.

The rugged fuel capsule is contained within a graphite fuel cask from launch through lunar landing. The cask is designed to provide reentry heating protection and added containment for the fuel capsule in the event of an aborted mission. The cylindrical cask with hemispherical ends includes a primary graphite heat shield, a secondary beryllium thermal shield, and a fuel capsule support structure made of titanium and Inconel materials. The cask is 23 inches long and eight inches in diameter and weighs about 24.5 pounds. With the fuel capsule installed, it weighs about 40 pounds. It is mounted on the lunar module descent stage by a titanium support structure.

Once the lunar module is on the Moon, an Apollo astronaut will remove the fuel capsule from the cask and insert it into the SNAP-27 generator which will have been placed on the lunar surface near the module.

The spontaneous radioactive decay of the plutonium-238 within the fuel capsule generates heat in the generator. An assembly of 442 lead telluride thermoelectric elements converts this heat -- 1480 thermal watts -- directly into electrical energy -- at least 63.5 watts. There are no moving parts.

The unique properties of plutonium-238 make it an excellent isotope for use in space nuclear generators. At the end of almost 90 years, plutonium-238 is still supplying half of its original heat. In the decay process, plutonium-238 emits mainly the nuclei of helium (alpha radiation), a very mild type of radiation with a short emission range.

Before the use of the SNAP-27 system in the Apollo program was authorized, a thorough review was conducted to assure the health and safety of personnel involved in the launch and of the general public. Extensive safety analyses and tests were conducted which demonstrated that the fuel would be safely contained under almost all credible accident conditons.

Lunar Orbital Science

One of the primary mission objectives, that of photographing candidate exploration sites, will be accomplished in lunar orbit. In addition, several experiments, photographic and scientific tasks will be carried out in lunar orbit.

CM PHOTOGRAPHIC TASKS

● PRIMARY OBJECTIVE
  ● PHOTOGRAPH CANDIDATE EXPLORATION SITE: DESCARTES

● SECONDARY OBJECTIVES
  ● LUNAR SURFACE

    AREAS OF SCIENTIFIC INTEREST

    MAN-MADE OBJECTIVES

      APOLLO 14 "LANDING LM"
      APOLLO 12 LM A/S IMPACT
      APOLLO 13 S-IVB IMPACT

      APOLLO 14 S-IVB IMPACT (SITE DEPENDENT)
      APOLLO 12 LM SITE (INCLUDED IN STRIP PHOTOGRAPHY)

● ASTRONOMIC:

    GALACTIC LIGHT   ZODIACAL LIGHT

    LUNAR LIBRATIONS (L_4)   EARTH DARKSIDE

      (TARGETS OF OPPORTUNITY)

The experiments include an S-band Transponder Test to determine variations in lunar gravity; a Bistatic Radar Experiment to determine electromagnetic reflective properties of the lunar surface; an Apollo Window Meteoroid Test to determine the effect of space particles on surfaces; and a Gegenschein from Lunar Orbit Experiment.

Photographic and scientific tasks include the photographing of lunar surface areas of high scientific interest; transearth lunar photos; investigations into visibility at high Sun angles; the updating of selendetic reference points; and photography of dim-light phenomena such as the Zodiacal light.

Photographs of a Candidate Exploration Site: The southern highland north of the crater Descartes is one of several sites under consideration for future long-stay Apollo missions. The Descartes area is characterized by hill, groove and furrow deposits in a highland basin, and is scientifically interesting from the standpoint of determining the age and composition of highland surface material as well as estimating volcanism time spans and compositional trends. The area is centered at eight degrees 51 minutes South Latitude by 15 degrees 34 minutes East Longitude.

Descartes will be photographed in several formats from low altitude shortly after the CSM/LM go into a 10x58nm lunar orbit with the descent orbit insertion maneuver. The lunar topographic camera mounted in the CM hatch window, a 70mm Hasselblad camera mounted in the right-hand rendezvous window, and a 16mm data acquisition camera on a sextant adapter will photograph Descartes simultaneously. The topographic camera compensates for spacecraft motion and makes overlapping high resolution stereo photographs. Descartes will again be photographed from 60nm after command module orbit circularization.

Visibility at High Sun Angles: This investigation is aimed at determining whether landing site visibility would be adequate for a safe landing if Sun angles were greater than 14 degrees, as would be in a launch delay until the T+24 hour opportunity.

Photography will supplement real-time crew observations of contrast and visibility of surface features under Sun angles ranging from 18 to 30 degrees providing additional data for planning future landing missions in which a delayed launch might be a factor. The 70mm photographs made during the observation runs will document crew voice descriptions for later comparisons and provide a basis for predictions in future missions.

Selenodetic Reference Point Update: This is a landmark tracking task to provide additional data for pinning down the exact location of selected lunar reference points. Map makers are hampered by inconsistencies in the lunar coordinates of surface features between photomosaics from manned and unmanned lunar missions and those shown on lunar aeronautical charts. A similar task was accomplished on the Apollo 12 mission, and the landmark tracking conducted by the Apollo 14 crew will further expand the net of pinpointed lunar surface features.

The CSM scanning telescope is used for tracking each landmark for about a three-minute period. At least four marks will be entered into the spacecraft computer on each landmark. The spacecraft computer, operating in Program 22 Orbital Navigation, integrates spacecraft attitude, optics shaft/trunnion angles and other factors to compute the actual lunar latitude and longitude of the landmark by postflight comparison with the spacecraft trajectory.

CSM Orbital Science Photography: Lunar surface areas and features of interest to scientists will be photographed by the command module pilot during the period he is alone in the command module. The lunar topographic camera, loaded with high-resolution black and white film and shooting 60 per cent overlapping photos, and two Hasselblad cameras--one using color film, the other using black and white--will be aimed at specific areas of interest. Additionally, certain lunar areas in Earthshine will be photographed.

These photographs, with those obtained during Apollos 8 and 12 and from the Ranger, Surveyor, and Lunar Orbiter Programs, will help answer questions about the Moon, generate new questions as new data are revealed, guide future mission planning, and allow for extrapolation of "ground truth" data collected at the landing sites to larger segments of the lunar surface.

Transearth Lunar Photography: Much of the photography as well as Earth observations of the Moon's surface features have been only of the side visible from Earth. Shortly after transearth injection on Apollo 14, the lunar topographic camera and two Hasselblads will be used to gather photographic data of the lunar farside and the eastern limb of the Moon. The photos will be correlated with existing orbital stereo photos and landmark tracking data for determining the positions of lunar features at latitudes not flown over in near-equatorial Apollo lunar orbits. The transearth lunar photography task will result in an increase in knowledge of the size, shape and mass distribution of the Moon as well as provide a source of data for improving lunar surface charts.

Dim Light Photography: Low-brightness light sources in the depths of the celestial sphere will be the targets for the dim light photography task. A 16mm data acquisition camera loaded with a high speed black and white film rated at ASA 6,000 will be the tool used to photograph these sources.

Among the subjects to be photographed will be the north pole of the galaxy of which our solar system is a part, the north ecliptic pole, the north pole of the celestial sphere, and the northernmost portion of the Milky Way; zodiacal light levels in the periods just prior to CSM lunar sunrise; the lunar libration region L4; Earth limb during solar eclipse by the Earth; comets (celestial conditions permitting); and the Earth's darkside as photographed through the command module sextant.

Even with the high-speed film, exposure times will run as long as 60 seconds, thus requiring a stable spacecraft attitude during photography sessions.

Zodiacal light is a faint glow extending around the entire zodiac, but most noticeable in the region of the Sun. The glow is theorized to be sunlight reflected from meteoritic-size particles in or near the ecliptic in the planetoid belt. Lunar libration region L4, located at 10 hrs 10 min right ascension and +10 degrees declination, is near the constella-tion Virgo as viewed from Apollo 14. It is a point in space on the trailing side of the Moon's orbital path, 60 degrees offset from both the Earth and Moon, at which astronomers conjecture that the gravitational potential is minimal and where particles of matter would tend to remain as a light reflective source.

Gegenschein from Lunar Orbit: This experiment is similar to the dim light photography task, and involves long exposures with the 16mm data acquisition camera on the high speed black and white film. All photos must be made while the command module is in total darkness in lunar orbit.

Gegenschein is a faint light source covering a 20-degree field of view along the Earth-Sun line on the opposite side of the Earth from the Sun (anti-solar axis). One theory on the origin of Gegenschein is that particles of matter are trapped at the Moulton Point and reflect sunlight. Moulton Point is a theoretical point located 940,000 statute miles from the Earth along the anti-solar axis where the sum of all gravitational forces is zero. From lunar orbit, the Moulton Point region can be photographed from about 15 degrees off the Earth-Sun axis, and the photos should show whether Gegenschein results from the Moulton Point theory or stems from zodiacal light or from some other source.

During the same time period that photographs of the Gegenschein and the Moulton Point are taken, photographs of the same regions will be obtained from the Earth. The principal investigator is Lawrence Dunkelman of the Goddard Space Flight Center.

S-Band Transponder: The objective of this experiment is to detect variations in lunar gravity along the lunar surface track. These anomalies in gravity result in minute perturbations of the spacecraft motion and are indicative of magnitude and location of mass concentrations on the Moon. The Manned Space Flight Network (MSFN) and the Deep Space Network (DSN) will obtain and record S-band Doppler tracking measurements of the docked CSM/LM and the undocked CSM while in lunar orbit; S-band Doppler tracking measurements of the LM during non-powered portions of the lunar descent; and S-band Doppler tracking measurements of the LM ascent stage during non-powered portions of the descent for lunar impact. The CSM and LM S-band transponders will be operated during the experiment period.

S-band Doppler tracking data have been analyzed from the Lunar Orbiter missions and definite gravity variations were detected. These results showed the existence of mass concentrations (mascons) in the ringed maria. Confirmation of these results has been obtained with Apollo tracking data.

With appropriate spacecraft orbital geometry much more
scientific information can be gathered on the lunar gravitational
field. The CSM and/or LM in low-altitude orbits can provide
new detailed information on local gravity anomalies. These data
can also be used in conjunction with high-altitude data to
possibly provide some description on the size and shape of the
perturbing masses. Correlation of these data with photographic
and other scientific records will give a more complete picture
of the lunar environment and support future lunar activities.
Inclusion of these results is pertinent to any theory of the
origin of the Moon and the study of the lunar subsurface structure.
There is also the additional benefit of obtaining better naviga-
tional capabilities for future lunar missions in that an improved
gravity model will be known. William Sjogren, Jet Propulsion
Laboratory, Pasadena, Calif., is principal investigator.

Bistatic Radar Experiment: The downlink Bistatic Radar
Experiment seeks to measure the electromagnetic properties of
the lunar surface by monitoring that portion of the spacecraft
telemetry and communications beacons which are reflected from
the Moon.

The CSM S-band telemetry beacon (f = 2.2875 Gigahertz),
the VHF voice communications link (f = 259.7 megahertz), and
the spacecraft omni-directional and high gain antennas are
used in the experiment. The spacecraft is oriented so that the
radio beacon is incident on the lunar surface and is successively
reoriented so that the angle at which the signal intersects
the lunar surface is varied. The radio signal is reflected
from the surface and is monitored on Earth. The strength of
the reflected signal will vary as the angle at which it intersects the surface is varied.

By measuring the reflected signal strength as a function
of angle of incidence on the lunar surface, the electromagnetic
properties of the surface can be determined. The angle at which
the reflected signal strength is a minimum is known as the
Brewster Angle and determines the dielectric constant. The
reflected signals can also be analyzed for data on lunar surface
roughness and surface electrical conductivity.

The S-band signal will primarily provide data on the surface.
However, the VHF signal is expected to penetrate the gardened
debris layer (regolith) of the Moon and be reflected from the
underlying rock strata. The reflected VHF signal will then
provide information on the depth of the regolith over the Moon.

The S-band BRE signal will be monitored by the 210-foot antenna at the Goldstone, Calif., site and the VHF portion of the BRE signal will be monitored by the 150-foot antenna at the Stanford Research Institute in California.

Lunar Bistatic Radar Experiments were also performed using the telemetry beacons from the unmanned Lunar Orbiter I in 1966 and from Explorer 35 in 1967. Taylor Howard, Stanford University, is the principal investigator.

Engineering/Operational Objectives

Apollo 14 engineering and operational tasks and experiments include evaluation of the Modularized Equipment Transporter (MET), measurements of extravehicular mobility unit water use, command module oxygen flow rate measurements to evaluate oxygen tank performance, evaluations of the EVA communications system, and evaluation of different types of thermal coatings exposed to lunar dust.

CSM Oxygen Flow: The ability of the service module cryogenic oxygen tankage to deliver oxygen at a high flow rate, such as will be required during inflight EVA in the future Apollo missions, will be verified during Apollo 14.

Late in transearth coast the tank's delivery capability will be evaluated at low quantities in a simulated situation wherein two of the three tanks had failed. The high flow rate test and the low quantity test will be run separately and at different periods during transearth coast.

Modularized Equipment Transporter Evaluation: Ease of deployment from the lunar module equipment bay, wheel traction characteristics and crew handling and utility of the MET will be evaluated and photographed during the periods the MET is used. The amount of force needed to move the vehicle over various types of lunar surface will be estimated.

EMU Water Consumption Measurement: The quantity of water remaining in the crew portable life support system (PLSS) backpacks presently is calculated on the basis of estimated metabolic heat load from biomedical telemetry correlated with PLSS feedwater usage rate. In Apollo 14, a method will be tested through which a more accurate estimate can be made of water quantity remaining.

At the end of EVA 1, after the commander and lunar module pilot have repressurized the lunar module, water remaining in one PLSS will be drained into a bag and weighed. The PLSS remote control unit will be weighed for establishing the lunar weight to the known Earth weight to establish a factor for water weight calculation.

Thermal Coating Degradation: The effects of lunar soil upon the optical properties of twelve various types of thermal coatings will be measured to gather data for use on future lunar surface equipment. The tests will be made during EVA 2. Two thermal control sample holders, each fitted with identical sample swatches will be used--one for control, and the other coated with lunar dust which is then brushed off. Stereo photographs will be made at each stage of the test as an adjunct to subjective crew evaluations of optical properties, (absorptivity and emissivity) of each type of material after the lunar dust has been brushed away.

Coating samples include six types of white paint, vacuum-deposited silver on quartz, silver/Inconel on Teflon, aluminum on kapton, oxidized silicon monoxide on aluminized kapton, anodized 6,061 aluminum and white dacron fabric on aluminized mylar.

EVA Communication System Performance: This task will determine the ability of the PLSS communication system signals to reach the lunar module antenna for relay of voice and data to Earth when an EVA crewman is behind terrain obstruction. Degradation of loss of communications when the commander is out of line-of-sight to the LM will be evaluated by the ground and by the LM pilot who will describe and photograph the relationship between the commander, the obstacle and the LM location.

Apollo Window Meteoroid (S-176): This experiment will be accomplished to determine the meteoroid cratering flux of particles responsible for degradation of surfaces exposed to the space environment. The command module windows will be inspected prior to launch and after return. Meteoroid craters in the command module windows will be analyzed in detail to aid in the development of space materials and prediction of material life-time in space. The principal investigator is Burton Cour-Palais of the Manned Spacecraft Center, Houston.

APOLLO LUNAR HAND TOOLS

Special Environmental Container - The special environmental sample is collected in a carefully selected area and sealed in a special container which will retain a high vacuum. The container is opened in the Lunar Receiving Laboratory where it will provide scientists the opportunity to study lunar material in its original environment.

Extension handle - This tool is of aluminum alloy tubing with a malleable stainless steel cap designed to be used as an anvil surface. The handle is designed to be used as an extension for several other tools and to permit their use without requiring the astronaut to kneel or bend down. The handle is approximately 24 inches long and one inch in diameter. The handle contains the female half of a quick disconnect fitting designed to resist compression, tension, torsion, or a combination of these loads.

Six core tubes - These tubes are designed to be driven or augered into loose gravel, sandy material, or into soft rock such as feather rock or pumice. They are about 15 inches in length and one inch in diameter and are made of aluminum tubing. Each tube is supplied with a removable non-serrated cutting edge and a screw-on cap incorporating a metal-to-metal crush seal which replaces the cutting edge. The upper end of each tube is sealed and designed to be used with the extension handle or as an anvil. Incorporated into each tube is a spring device to retain loose materials in the tube.

Scoops (large and small) - These tools are designed for use as a trowel and as a chisel. The scoop is fabricated primarily of aluminum with a hardened-steel cutting edge riveted on and a nine-inch handle. A malleable stainless steel anvil is on the end of the handle. The angle between the scoop pan and the handle allows a compromise for the dual use. The scoop is used either by itself or with the extension handle.

Sampling hammer - This tool serves three functions, as a sampling hammer, as a pick or mattock, and as a hammer to drive the core tubes or scoop. The head has a small hammer face on one end, a broad horizontal blade on the other, and large hammering flats on the sides. The handle is 14 inches long and is made of formed tubular aluminum. The hammer has on its lower end a quick-disconnect to allow attachment to the extension handle for use as a hoe. The head weight has been increased to provide more impact force.

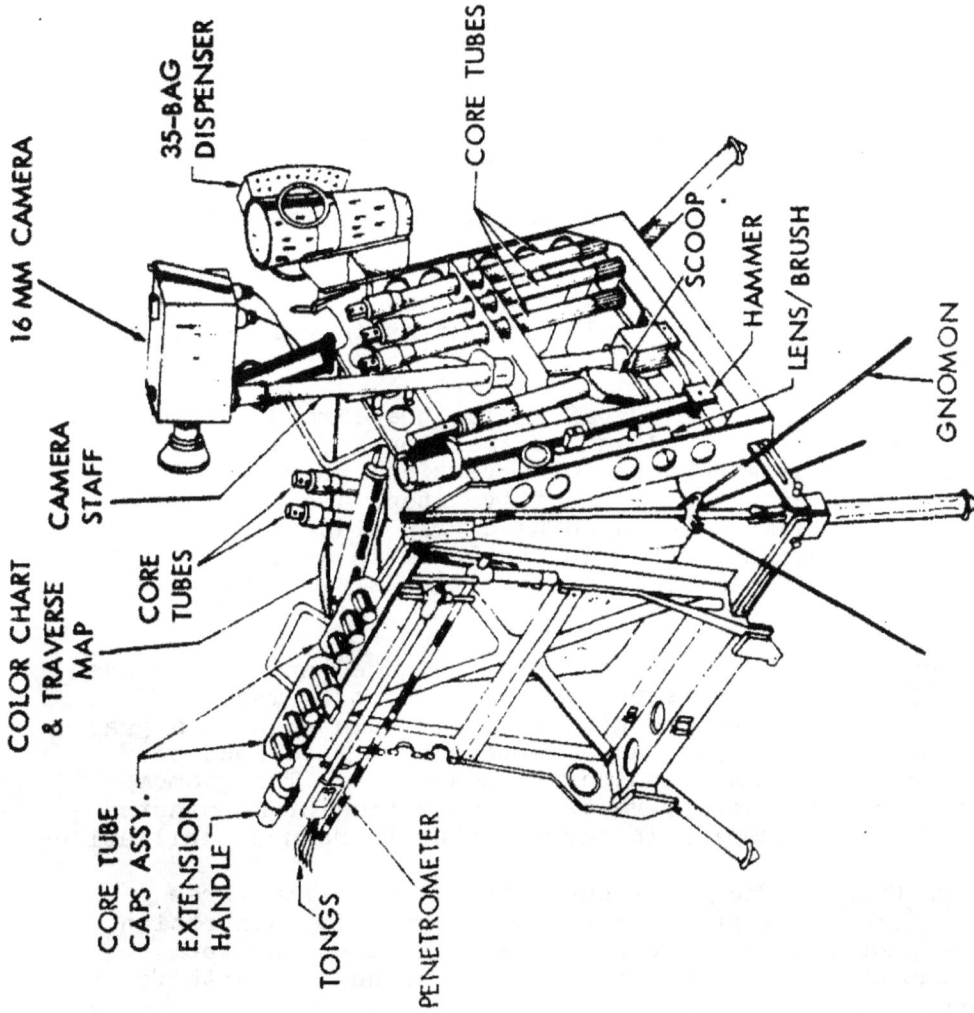

COLOR CHART & TRAVERSE MAP

16 MM CAMERA

35-BAG DISPENSER

CAMERA STAFF

CORE TUBES

CORE TUBE CAPS ASSY.

CORE TUBES

EXTENSION HANDLE

SCOOP

HAMMER

LENS/BRUSH

TONGS

GNOMON

PENETROMETER

APOLLO LUNAR HAND TOOL CARRIER (ALHT) MET TRAVERSE CONFIGURATION

Tongs - The tongs are designed to allow the astronaut
to retrieve small samples from the lunar surface while in a
standing position. The tines are of such angles, length,
and number to allow samples of from 3/8 up to 2-1/2-inch
diameter to be picked up. One tool is 24 inches in overall
length and the other is 32 inches.

Brush/Scriber/Hand Lens - A composite tool

(1) Brush - To clean samples prior to selection
(2) Scriber - To scratch samples for selection and to
     mark for identification
(3) Hand lens - Magnifying glass to facilitate sample
     selection

Spring Scale - To weigh two rock boxes and other bags
containing lunar material samples, to maintain weight budget
for return to Earth.

Instrument staff - The staff holds the 16mm data acquisition
camera. The staff breaks down into sections. The upper section
telescopes to allow generation of a vertical stereoscopic base
of one foot for photography. Positive stops are provided at
the extreme of travel. A-shaped hand grip aids in aiming and
carrying. The bottom section is available in several lengths
to suit the staff to astronauts of varying sizes. The device
is fabricated from tubular aluminum.

Gnomon - This tool consists of a weighted staff suspended
on a two-ring gimbal and supported by a tripod. The staff
extends 12 inches above the gimbal and is painted with a gray
scale. The gnomon is used as a photographic reference to
indicate local vertical, sun angle, and scale. The gnomon
has a required accuracy of vertical indication of 20 minutes
of arc. Magnetic damping is incorporated to reduce oscillations.

Color Chart - The color chart is painted with three
primary colors and a gray scale. It is used in calibration
for lunar photography. The scale is mounted on the tool
carrier but may easily be removed and returned to Earth for
reference. The color chart is six inches in size.

Tool Carrier - The carrier is the stowage container for the tools during the lunar flight. After the landing the carrier serves as a support for the sample bags and samples, and as a tripod base for the instrument staff. The carrier folds flat for stowage. For field use it opens into a triangular configuration. The carrier is constructed of formed sheet metal and approximates a truss structure. Six-inch legs extend from the carrier to elevate the carrying handle sufficiently to be easily grasped by the astronaut. The tool carrier is mounted on the MET for geology traverses.

Field Sample Bags - Approximately 35 bags four inches by five inches are included in the Apollo lunar hand tools for the packaging of samples. These bags are fabricated from Teflon FEP, and are in a dispenser on the hand tool carrier.

Trenching Tool - A trenching tool with a pivoting scoop has been provided for digging the two-foot deep soil mechanics investigation trench. The two-piece handle is five feet long. The scoop is eight inches long and five inches wide and pivots from in-line with the handle to 90°--similar to the trenching tool carried on infantry backpacks. The trenching tool is stowed in the MESA rather than in the tool carrier.

FRA MAURO LANDING SITE

The Apollo 14 landing site is located at 3° 40' 19" south latitude by 17° 27' 46" west longitude, about 30 miles north of the Fra Mauro crater--the same site selected for the aborted Apollo 13 mission.

The hilly region has been designated the Fra Mauro formation, a widespread geological area covering large portions of the lunar surface around Mare Imbrium (Sea of Rains). The 700-mile wide Mare Imbrium is the largest recognizable impact structure on the Moon, and is thought to have been formed by a major impact of a huge mass colliding with the Moon during the period when the Earth and the planets were forming. The Fra Mauro formation is believed to be made up of an ejecta blanket thrown out by that impact.

The area is characterized by ridges a few hundred feet high which radiate from the Imbrium basin separated by undulating valleys. The ejecta blanket now is buried by younger rubble and lunar soil churned up by more recent meteorite impacts and possibly moonquakes.

Fra Mauro debris may have come from as deep as 100 miles below the original lunar crust, and returned samples should show when the Imbrium basin was formed and help to establish the age and physical/chemical nature of pre-impact material from deep in the crust. It is theorized that Fra Mauro rocks will predate the Apollo 11 rocks (4.6 billion years) and the Apollo 12 rocks (3.5 billion years) to a period near the original age of the Moon.

A recent impact near the landing point formed Cone crater, nearly 1,000 feet across and 250 feet deep, with large blocks of original Imbrium material around the crater rim. Shepard and Mitchell will climb Cone crater's gently-sloping outer wall to photograph the crater's interior and chip samples from the boulders around the edge.

The Fra Mauro formation became more interesting to scientists when the Apollo 12 seismometer at Surveyor crater 110 miles to the west relayed to Earth signals of monthly moonquakes believed to have originated in the Fra Mauro crater as the Moon passed through its perigee.

The Fra Mauro crater and surrounding formation take
their names from a 15th century Italian monk and mapmaker,
who in 1457 mapped the then-known Mediterranean world with
surprising accuracy.

PHOTOGRAPHIC EQUIPMENT

Still and motion pictures will be made of most space-craft maneuvers and crew lunar surface activities. During lunar surface activities, emphasis will be on photographic documentation of lunar surface features and lunar material sample collection. From orbital altitude, photographic tasks and experiments will include high resolution photography to support future landing missions, photographs of lunar surface areas of scientific interest, and astronomical photography of the Gegenschein zodiacal light, libration points, galactic poles, and the earth's dark side.

Camera equipment stowed in the Apollo 14 command module consists of two 70mm Hasselblad electric cameras, a 16mm motion picture camera, and the Hycon lunar topographic camera (LTC).

The LTC is stowed beneath the commander's couch. In use, the camera mounts in the crew access hatch window.

The LTC with 18-inch focal length f/4.0 lens provides resolution of objects as small as 15-25 feet from a 60-nm altitude and as small as three to five feet from the 8-nm pericynthion. Film format is 4.5-inch square frames on 100 foot long rolls, with a frame rate variable from four to 75 frames a minute. Shutter speeds are 1/50, 1/100, and 1/200 second. Spacecraft forward motion during exposures is compensated for by a servo-controlled rocking mount. The film is held flat in the focal plane by a vacuum platen connected to the auxiliary dump valve.

The camera weighs 65 pounds without film, is 28 inches long, 10.5 inches wide, and 12.25 inches high. It is a modification of an aerial reconnaissance camera.

Cameras stowed in the lunar module are two 70mm Hasselblad data cameras fitted with 60mm Zeiss Metric lenses, two 16mm motion picture cameras fitted with 10mm lenses, and a Kodak closeup stereo camera for high resolution photos on the lunar surface. The LM Hasselblads have crew chest mounts that leave both hands free.

-more-

One of the command module Hasselblad electric cameras is normally fitted with an 80mm f/2.8 Zeiss Planar lens, but a bayonet mount 250 mm lens may be substituted for special tasks.

The second Hasselblad camera is fitted with an 80mm lens and a Reseau plate which allows greater dimensional control on photographs of the lunar surface.

The 80mm lens has a focussing range from three feet to infinity and has a field of view of 38 degrees vertical and horizontal on the square-format film frame. Accessories for the command module Hasselblads include a spotmeter, intervalometer, remote control cable, and film magazines. Hasselblad shutter speeds range from time exposure and one second to 1/500 second.

The Maurer 16mm motion picture camera in the command module has lenses of 10, 18, and 75mm available. The camera weighs 2.8 pounds with a 130-foot film magazine attached. Accessories include a right-angle mirror, a power cable, and a sextant adapter which allows the camera to use the navigation sextant optical system. One of the LM motion picture cameras will be mounted in the right-hand window to record descent and landing and the two EVA periods and later will be taken to the surface.

The 35mm stereo closeup camera stowed in the LM MESA shoots 24mm square stereo pairs with an image scale of one-half actual size. The camera is fixed focus and is equipped with a stand-off hood to position the camera at the proper focus distance. A long handle permits an EVA crewman to position the camera without stooping for surface object photography. Detail as small as 40 microns can be recorded. The camera allows photography of significant surface structure which would remain intact only in the lunar environment, such as fine powdery deposits, cracks or holes, and adhesion of particles. A battery-powered electronic flash provides illumination, and film capacity is a minimum of 100 stereo pairs.

TELEVISION

Apollo 14 will carry two color and one black-and-white television cameras. One color camera will be used for command module cabin interiors and out-the-window Earth/Moon telecasts, and the other color camera will be stowed in the LM descent stage from where it will view the astronaut initiate egress to the lunar surface and later will be deployed on a tripod to transmit a real-time picture of the two periods of lunar surface EVA.

The two color TV cameras are identical except for additional thermal protection on the lunar surface camera. Built by Westinghouse Electric Corp., Aerospace Division, Baltimore, Md., the color cameras put out a standard 525-line, 30 frame-per-second signal in color by use of a rotating color wheel system.

The color TV cameras weigh 12 pounds and are fitted with zoom lenses for wide-angle or closeup fields of view. The CM camera is fitted with a three-inch monitor for framing and focusing. The lunar surface color camera has 100 feet of cable available.

The backup black and white lunar surface TV camera, also built by Westinghouse, is of the same type used in the first manned lunar landing in Apollo 11. It weighs seven pounds and draws 6.5 watts of 24-32 volts direct current power. Scan rate is ten frames-per-second at 325 lines per frame. The camera body is 10.6 inches long, 6.5 inches wide and 3.4 inches deep, and is fitted with bayonet-mounted wide-angle and lunar day lenses.

During the two lunar surface EVA periods, the Apollo commander will be recognizable by red stripes around the elbows and knees of his pressure suit.

The following is a preliminary plan for TV transmissions based upon a 3:23 p.m. EST Jan. 31 launch.

Apollo 14 TV Schedule

DAY	DATE	EST	GET, HR:MIN	DURATION, HR:MIN	ACTIVITY SUBJECT	VEHICLE	STATION
Sunday	Jan. 31	6:28 PM	03:05	00:25	Transposition & Docking	CSM	GDS
Wednesday	Feb. 3	5:08 AM	61:45	00:45	Interior & IVT to LM	CSM	GDS/HSK
Thursday	Feb. 4	8:23 PM	101:00	00:14	Fra Mauro landing site	CSM	GDS
Friday	Feb. 5	9:20 AM	113:40	04:00	Lunar surface EVA-1	LM	HSK/MAD
Saturday	Feb. 6	4:59 AM	133:31	07:43	Lunar surface EVA-2	LM	HSK/GDS/MAD
Saturday	Feb. 6	3:14 PM	143:51	00:06	Rendezvous	CSM	MAD
Saturday	Feb. 6	3:29 PM	144:06	00:04	Docking	CSM	MAD
Sunday	Feb. 7	7:53 PM	172:30	00:30	Inflight demonstrations	CSM	GDS

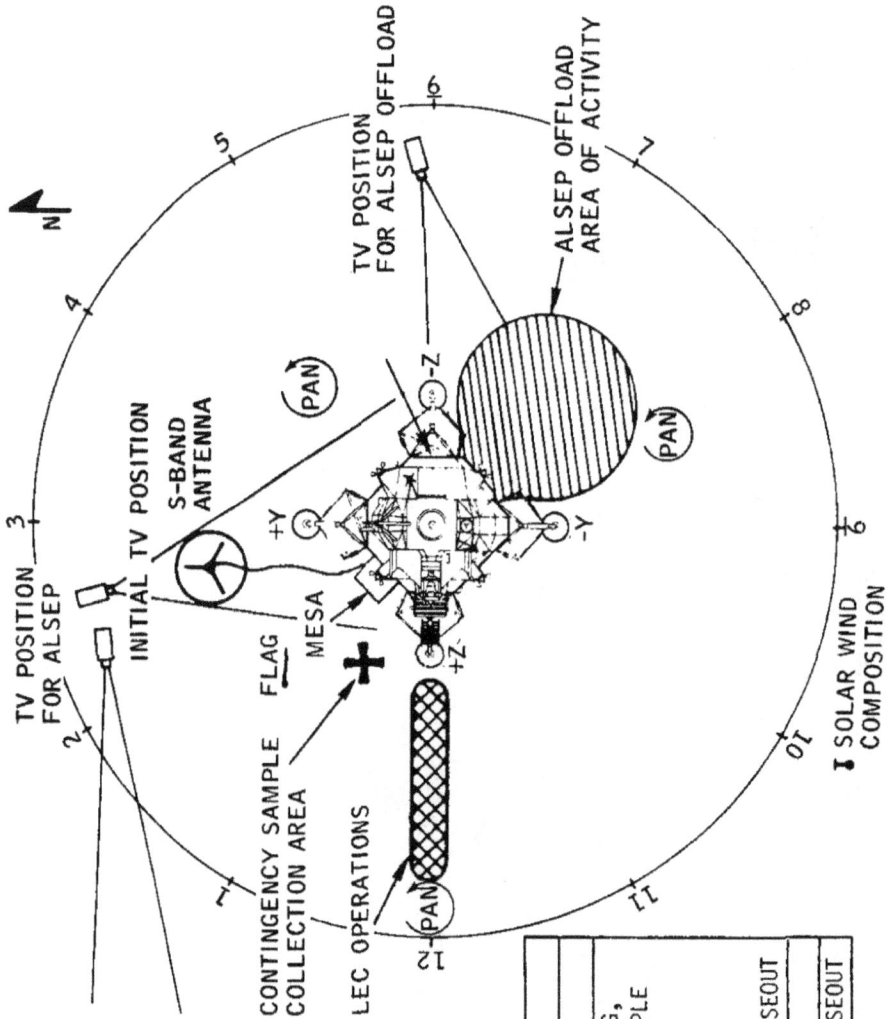

PROBABLE TV AREAS FOR NEAR LM SURFACE ACTIVITIES

TIME	POSITION	VIEW
EVA-1		
0+30	2:30 @ 50'	MESA, STEPS, FLAG, CONTINGENCY SAMPLE
1+07	SAME	PANORAMA
1+19	6:00 @ 30'	SEQ BAY
1+33	2:30 @ 50'	ALSEP TRAVERSE
3+45	SAME	MESA, LADDER, CLOSEOUT
EVA-2		
CONT	2:30 @ 50'	MESA, LADDER, CLOSEOUT

ZERO-GRAVITY INFLIGHT DEMONSTRATIONS

Four "Zero-Gravity Inflight Demonstrations" are being flown on Apollo 14. Zero-Gravity Inflight Demonstrations are technical demonstrations of equipment and processes designed to illustrate the utilization of the unique condition of zero-gravity in space. The tests are planned during the relatively inactive return to Earth phase of the mission and are to be performed at the option of the crew.

These technical demonstrations result from NASA studies conducted at the Marshall Space Flight Center, Huntsville, Ala., and the Lewis Research Center, Cleveland. They are simple tests that could provide information on zero-gravity effects useful in supporting the establishment of design requirements for future experiments in the Materials Science and Manufacturing in Space program.

Each demonstration is stowed in the Apollo 14 command module. The units require only a small amount of power from the spacecraft for operation and lighting. Operation of the demonstrations essentially requires only activation of the tests by the astronauts. Data will be obtained by crew observations and photography during the mission and laboratory tests following the mission.

The four technical demonstrations planned for Apollo 14 are:

1. Electrophoretic Separation

2. Heat Flow and Convection

3. Liquid Transfer

4. Composite Casting

Electrophoretic Separation

Most organic molecules pick up small electric charges when they are placed in slightly acid or alkaline water solutions and will move through such a solution if an electric field is applied to it; this effect is known as electrophoresis. Since different molecules move at different speeds, the faster molecules in a mixture that starts moving from one end of a tube of solution will outrun the slower ones as they move toward the other end. This characteristic of electrophoresis can be exploited to prepare pure samples of organic materials for applications in medicine and biological research if problems due to sample sedimentation and sample mixing by convection can be overcome.

The electrophoretic separation demonstration is designed to test an engineering approach to performing the preparation process in space, where the weightlessness of the solutions and sample mixtures should suppress both convection and sedimentation. A small, specially designed electrophoretic separation apparatus will be tested and the quality of the separations obtained will be demonstrated by trails with three sample mixtures having widely different molecular weights: (1) a mixture of red and blue organic dyes; (2) human hemoglobin; and (3) DNA (the molecules that carry the genetic code) from salmon sperm.

The apparatus for the electrophoretic separation demonstration is simple, but capable of providing useful information on the development and use of electrophoresis equipment on future missions.

Hardware consists of three tubes containing, respectively, organic dye, hemoglobin and salmon DNA suspended in a diluted electrolyte solution. Other gear includes a pump and motor, gas-phase separator, fluorescent lamps, a power supply, and a bimetallic thermometer.

The equipment will weigh about nine pounds and be stored in a box seven by five by four inches. The power needed is 30 watts at 400 cycles AC. When in use, the demonstration will be strapped to a vacant space on a bulkhead.

The demonstration will take about one hour to complete. Photography will be the principal data recovery method, using a 70 mm onboard camera with color film. Minimum crew training will be necessary.

The three tubes will have electrodes mounted in their ends, separated from the samples by porous membranes. A voltage will be applied to the electrodes and the electrolyte will be circulated over them by the pump and motor.

Hydrogen and oxygen produced by the electrolysis of the solution will be collected and absorbed by the gas separators.

The fluorescent lamps will illuminate the sample tubes for photography. Visible light will be used to illuminate the dyes and hemoglobin, and ultraviolet light will be used for the DNA.

If successful, the demonstration will show that more refined apparatus could be developed to prepare samples of materials on future space missions for use in medical and biological research on the ground. Ultimately, the method may prove practical for large-scale processing of new vaccines and similar biological preparations on board manned space stations.

The General Electric Co. built the demonstration package under contract to MSFC.

Heat Flow and Convection

The Heat Flow and Convection demonstration is designed to perform four tests on heat transfer in weightless liquids and gases. In three of the tests temperatures around electric heaters immersed in samples of pure water, a sugar solution, and carbon dioxide gas will be mapped out by color changes produced in "liquid crystal" temperature indicators. The fourth test will observe the fluid flow induced by heating a sample of oil containing a suspension of fine aluminum flakes.

The demonstration is contained in a nine by nine by three and one-half-inch package weighing seven pounds. It requires 30 watts of power.

Hardware includes a control panel; two zone tubes containing fluids of different viscosity (distilled water and a sugar/water solution); one radial convection cell containing carbon dioxide; and a flow pattern cell with a fluid level indicator. The flow pattern cell is filled with Krytox, an oil used as a lubricant in the spacecraft. The fluids selected for this application were chosen because they will yield the required information and are safe to use in the spacecraft.

To operate the demonstration, one of the crewmen will mount the package below the command module's center couch. The 16 mm onboard camera will be mounted several inches from the package's front panel. After plugging the experiment and camera into the command module's 28 volt DC power source, the operator will turn on the heaters and camera.

The remaining procedure is more or less automatic. As the increasing temperatures cause the "liquid crystals" to change color, the camera will photograph the process at a rate of one frame-per-second for one hour and 28 minutes. The astronaut crewman will be asked to use only one magazine of film in this demonstration. However, a second magazine could be used if the crewman should decide to repeat the procedure. The film will be returned for evaluation at MSFC. The onboard television camera may also be used to record some of the demonstration.

The results observed and photographed by the astronauts will characterize the effects of convection and other modes of heat transfer in fluids during space flight. This information will be of value in designing future space experiments and assessing the feasibility of many processes that have been proposed for manufacturing products in space. Lockheed Missiles and Space Co. built the demonstration package.

Liquid Transfer

The Liquid Transfer technical demonstration is designed to demonstrate the benefits of utilizing tank baffling in the storage and transfer of liquids in zero-gravity. The tests will be conducted with two sets of simulated tanks, one set containing tank baffling and the other without any baffling. By observing and photographing the transfer of liquids in the two sets of tanks, a comparison can be made to determine the benefits obtained from the use of baffles in zero-gravity.

Experiment elements include the four tanks, a hand-actuated pump, and a light mounted behind the tanks. The tank assembly is six by ten inches by three inches. The pump is 2.5 inches in diameter by 6.5 inches. The experiment weighs nine pounds and uses 28 volts DC, 30 watts power.

The advantage of tanks with baffles can be important to the design considerations of future space refueling systems.

Composite Casting

This technical demonstration is designed to demonstrate the effect of zero-gravity on the preparation of cast metals, fiber-strengthened materials and single crystals. These test specimens will be processed in a small heating chamber in flight, for examination and testing upon return to Earth.

The demonstration will be done with low-melting-point (approximately 160° F) metal alloys and organic crystalline materials which will be models of metal matrix materials such as aluminum or nickel. Dispersants will include particles, chopped fibers or wires, whiskers and combinations of these materials with argon gas bubbles.

Demonstration hardware will include 18 hermetically-sealed aluminum capsules, containing composite material samples to be processed in space, a low-powered electrical resistance heater and a storage box, which is also a heat sink to cool the specimens.

The entire demonstration package will weigh about 12 pounds. Each of the sample containers is 3.5 inches long and 7/8-inch in diameter. The cylindrical heater unit is 4.25 by 5 by 3.5 inches in size, and has capped openings top and bottom. Power for the heater is provided by 28 volts DC.

An astronaut needs little training to do the experiment. He opens the storage box and secures it inside the command module tunnel with a spring retention clamp. He then removes a sample container from its stowage bag, inserts it in the heater and remounts the heater in the box. The heater inside the box is connected to a power cable and the unit is heated for about five minutes.

The man then takes the heater from the box and may shake the heater (like a cocktail shaker) for a few seconds to disperse particles better inside the sample. The heater's bottom cap is then opened and the heater is placed on a heat sink built into the storage box. After cooling for about 20 to 25 minutes, the sample container is removed and returned to its stowage bag. Another sample container can then be processed.

All 18 of the samples may not be used; the deciding factor is how much free time the astronauts have during their return trip from the Moon. No photography or television will be used in data collection. Basic data will be gained through post-flight evaluation of returned specimens, including metallurgical, chemical and mechanical properties tests.

The results to be obtained from these tests will be used to evaluate the prospects for making improved metallurgical products in space.

The hardware was built at NASA's Marshall Space Flight Center, Huntsville.

ASTRONAUTS AND CREW EQUIPMENT

Space Suits

Apollo crewmen will wear two versions of the Apollo space suit: an intravehicular pressure garment assembly worn by the command module pilot and the extravehicular pressure garment assembly worn by the commander and the lunar module pilot. Both suits are basically identical except that the extravehicular version has an integral thermal/meteoroid garment over the basic suit.

From the skin out, the basic pressure garment consists of a nomex comfort layer, a neoprene-coated nylon pressure bladder and a nylon restraint layer. The outer layers of the intravehicular suit are, from the inside out, nomex and two layers of Teflon-coated Beta cloth. The extravehicular integral thermal/meteoroid cover consists of a liner of two layers of neoprene-coated nylon, seven layers of Beta/Kapton spacer laminate, and an outer layer of Teflon-coated Beta fabric.

The extravehicular suit, together with a liquid cooling garment, portable life support system (PLSS), oxygen purge system, lunar extravehicular visor assembly and other components make up the extravehicular mobility unit (EMU). The EMU provides an extravehicular crewman with life support for a four-hour mission outside the lunar module without replenishing expendables. EMU total weight is 183 pounds. The intravehicular suit weighs 35.6 pounds.

Liquid cooling garment--A knitted nylon-spandex garment with a network of plastic tubing through which cooling water from the PLSS is circulated. It is worn next to the skin and replaces the constant-wear garment during EVA only.

Portable life support system--A backpack supplying oxygen at 3.9 psi and cooling water to the liquid cooling garment. Return oxygen is cleansed of solid and gas contaminants by a lithium hydroxide canister. The PLSS includes communications and telemetry equipment, displays and controls, and a main power supply. The PLSS is covered by a thermal insulation jacket. (Two stowed in LM).

Oxygen purge system--Mounted atop the PLSS, the oxygen purge system provides a contingency 45-minute supply of gaseous oxygen in two two-pound bottles pressurized to 5,880 psia. The system may also be worn separately on the front of the pressure garment assembly torso. It serves as a mount for the VHF antenna for the PLSS. (Two stowed in LM).

BACKPACK SUPPORT STRAPS

OXYGEN PURGE SYSTEM

LUNAR EXTRAVEHICULAR VISOR

BACKPACK CONTROL BOX

SUNGLASSES POCKET

OXYGEN PURGE SYSTEM ACTUATOR

PORTABLE LIFE SUPPORT SYSTEM

PENLIGHT POCKET

CONNECTOR COVER

COMMUNICATION, VENTILATION, AND LIQUID COOLING UMBILICALS

OXYGEN PURGE SYSTEM UMBILICAL

LM RESTRAINT RING

INTEGRATED THERMAL METEOROID GARMENT

EXTRAVEHICULAR GLOVE

UTILITY POCKET

URINE TRANSFER CONNECTOR, BIOMEDICAL INJECTION, DOSIMETER ACCESS FLAP AND DONNING LANYARD POCKET

LUNAR OVERSHOE

-more-

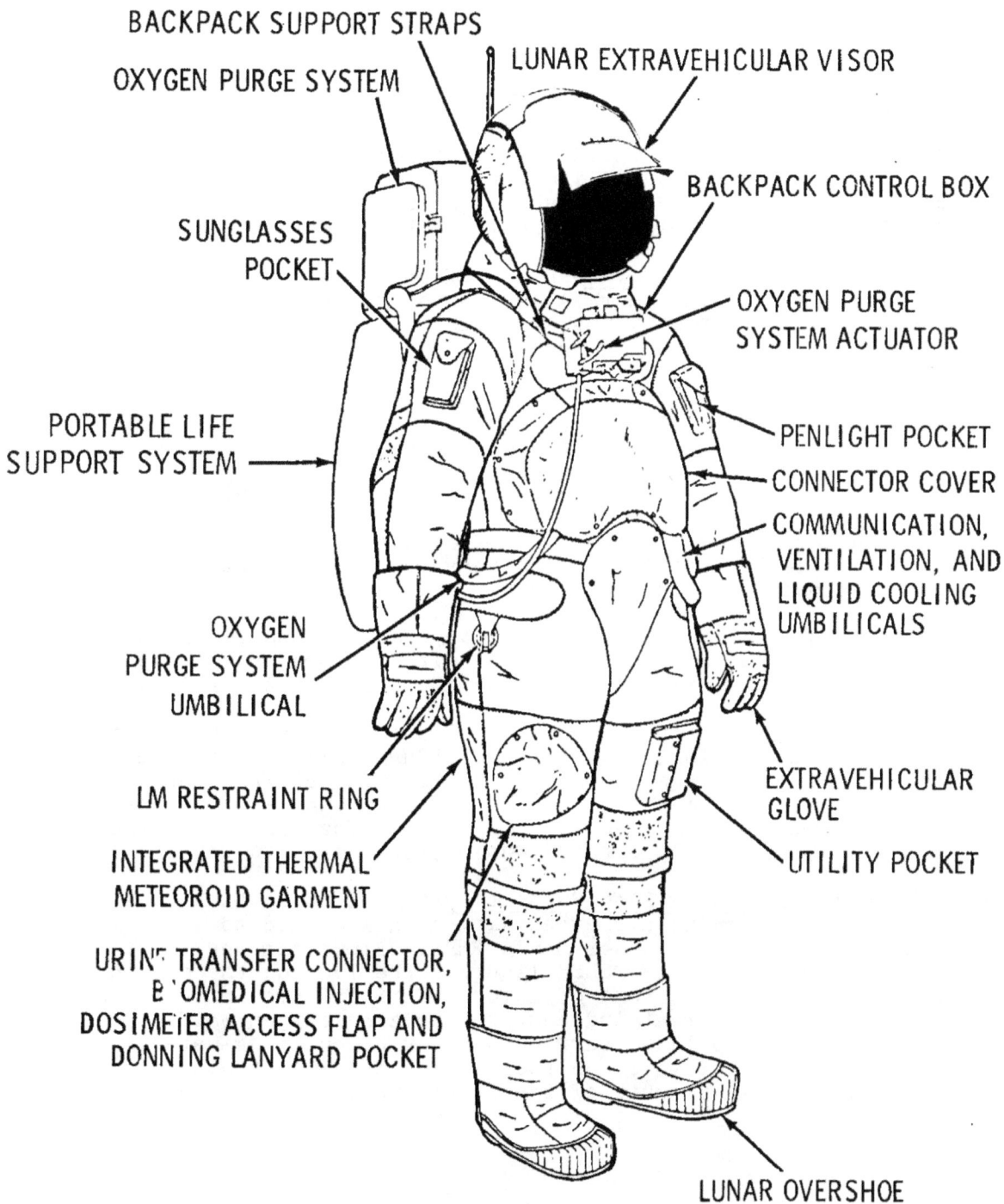

EXTRAVEHICULAR MOBILITY UNIT

Lunar extravehicular visor assembly--A polycarbonate
shell and two visors with thermal control and optical
coatings on them. The EVA visor is attached over the
pressure helmet to provide impact, micrometeoroid, thermal
and ultraviolet-infrared light protection to the EVA crewmen. Since Apollo 12, a sunshade has been added to the outer
portion of the LEVVA in the middle portion of the helmet rim.

Extravehicular gloves--Built of an outer shell of
Chromel-R fabric and thermal insulation to provide protection
when handling extremely hot and cold objects. The finger
tips are made of silicone rubber to provide more sensitivity.

A one-piece constant-wear garment, similar to "long
johns," is worn as an undergarment for the space suit in intravehicular operations and for the inflight coveralls. The
garment is porous-knit cotton with a waist-to-neck zipper for
donning. Biomedical harness attach points are provided.

During periods out of the space suits, crewmen wear
two-piece Teflon fabric inflight coveralls for warmth and for
pocket stowage of personal items.

Communications carriers ("Snoopy Hats") with redundant
microphones and earphones are worn with the pressure helmet;
a lightweight headset is worn with the inflight coveralls.

Another modification since Apollo 12 has been the addition
of eight-ounce drinking water bags ("Gunga Dins") attached to
the inside neck rings of the EVA suits. The crewman can take
a sip of water from the 6 by 8-inch bag through a 1/8-inch-
diameter tube within reach of his mouth. The bags are filled
from the lunar module potable water dispenser.

Buddy Secondary Life Support System--A connecting hose
system which permits a crewman with a failed PLSS to share
cooling water in the other man's PLSS. Flown for the first
time on Apollo 14, the BSLSS lightens the load on the oxygen
purge system in the event of a total PLSS failure in that the
OPS would supply breathing and pressurizing oxygen while the
metabolic heat would be carried away by circulating water.
The BSLSS will be stowed on the MET for EVA-2. It will not be
carried on EVA-1 because of the short distance from the LM.

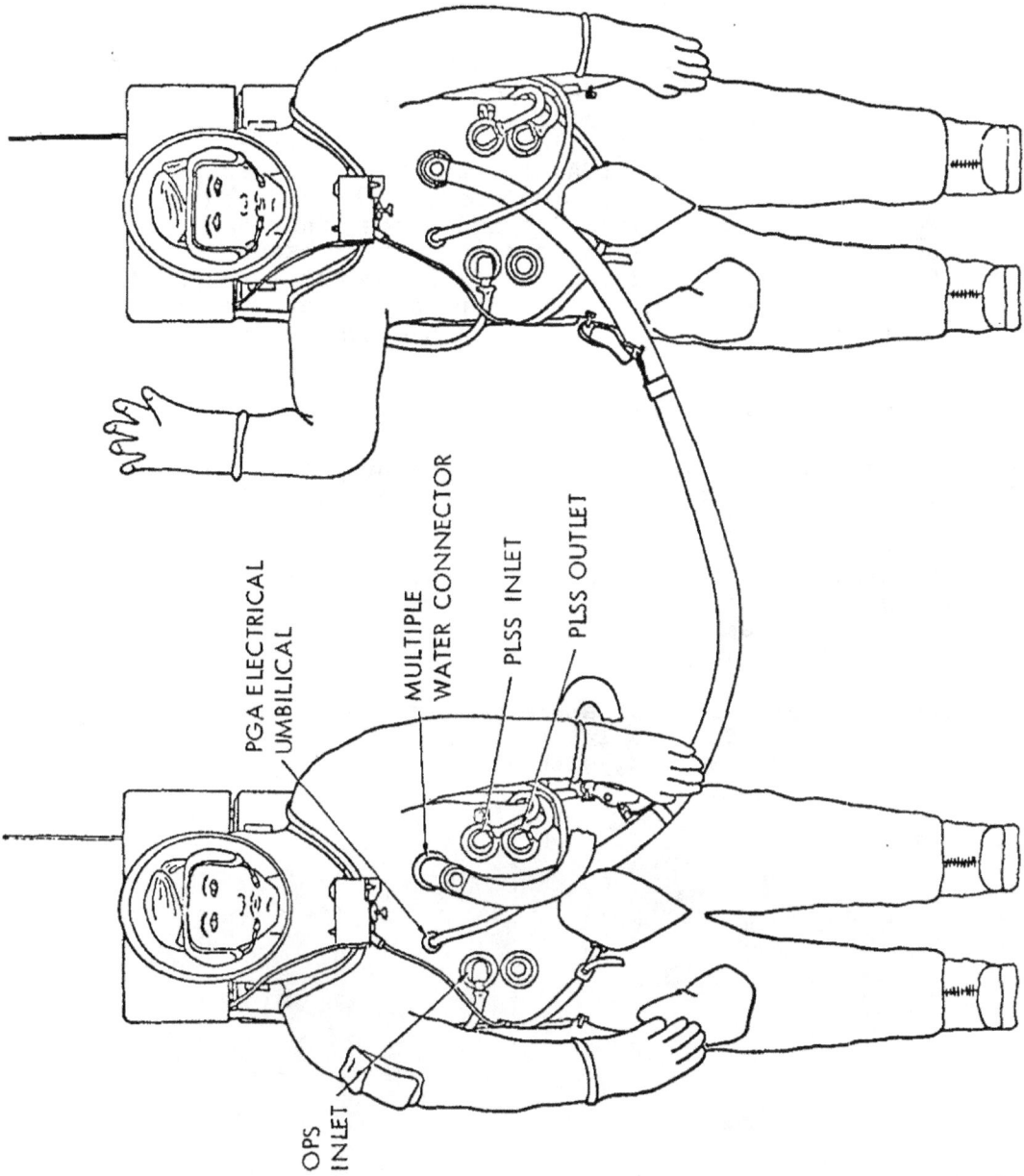

BUDDY SYSTEM FOR -6 PLSS

PGA ELECTRICAL
UMBILICAL

MULTIPLE
WATER CONNECTOR

PLSS INLET

PLSS OUTLET

OPS
INLET

Modularized Equipment Transporter--The MET is a two-wheeled vehicle with a tubular structure 86 inches long, 39 inches wide and 32 inches high when deployed ready for use on the lunar surface. The MET has a single handle for towing and has two legs to provide four-point stability at rest.

The MET is stowed during flight in the Modularized Equipment Storage Assembly (MESA) in the LM descent stage, and will be used during EVA-1. Equipment to be mounted on the MET for the geology traverse includes the lunar hand tool carrier and the geology tools it carries, the closeup stereo camera, two 70 mm Hasselblad cameras, a 16 mm data acquisition camera, film magazines, a dispenser for sample bags, a trenching tool, work table, sample weigh bags and the Lunar Portable Magnetometer.

The MET tires are 4 inches wide and 16 inches in diameter, and will be inflated with 1.5 psia nitrogen preflight. The tires will be baked at 250° F for 24 hours preflight to remove most of the antioxidants in the rubber. Operating limits for the MET tires are -70° F to +250° F.

Empty weight of the MET is 26 pounds, and the vehicle has a useful payload of about 140 pounds (Earth weight) including the lunar soil samples to be brought back to the LM from the geology traverse.

Estimated travel rate of a crewman towing the MET, as determined by tests with the 1/6-g centrifuge rig at MSC, is about 3.5 feet per second, with about one pound of pull required on level sand.

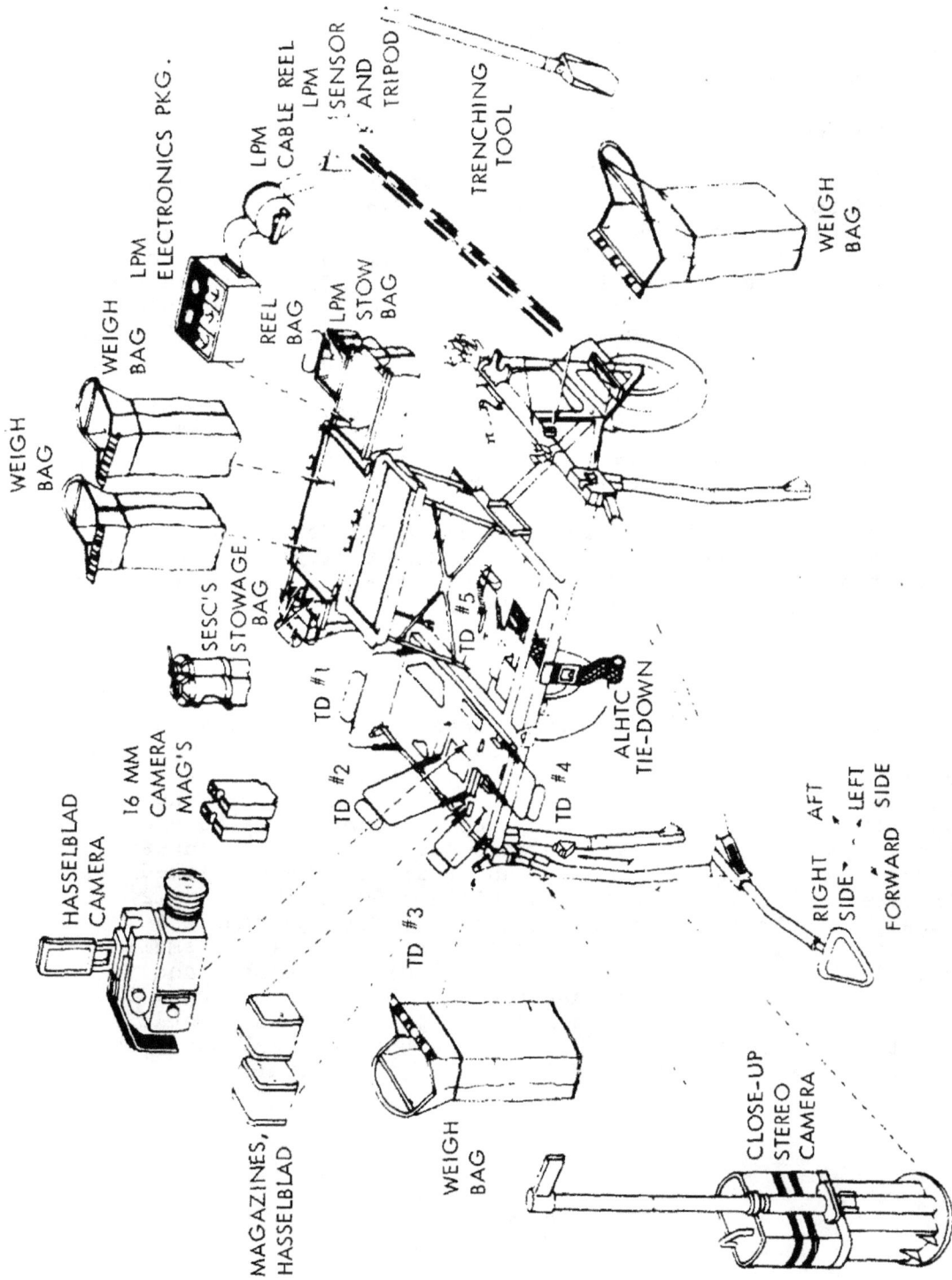

WEIGH BAG

LPM ELECTRONICS PKG.

LPM CABLE REEL

LPM SENSOR AND TRIPOD

TRENCHING TOOL

WEIGH BAG

REEL BAG

LPM STOW BAG

WEIGH BAG

WEIGH BAG

SESC'S STOWAGE BAG

TD #1

TD #5

TD #2

ALHTC TIE-DOWN

TD #4

HASSELBLAD CAMERA

16 MM CAMERA MAG'S

RIGHT SIDE

AFT

LEFT SIDE

FORWARD SIDE

TD #3

MAGAZINES, HASSELBLAD

WEIGH BAG

CLOSE-UP STEREO CAMERA

MET

-more-

Personal Hygiene

Crew personal hygiene equipment aboard Apollo 13
includes body cleanliness items, the waste management system
and one medical kit.

Packaged with the food are a toothbrush and a two-ounce
tube of toothpaste for each crewman. Each man-meal package
contains a 3.5-by-4-inch wet-wipe cleansing towel. Additionally, three packages of 12-by-12-inch dry towels are stowed
beneath the command module pilot's couch. Each package contains seven towels. Also stowed under the command module
pilot's couch are seven tissue dispensers containing 53 three-
ply tissues each.

Solid body wastes are collected in plastic defecation
bags which contain a germicide to prevent bacteria and gas
formation. The bags are sealed after use and stowed in empty
food containers for post-flight analysis.

Urine collection devices are provided for use while
wearing either the pressure suit or the inflight coveralls.
The urine is dumped overboard through the spacecraft urine dump
valve in the CM and stored in the LM.

Medical Kit

The 5 by 5 by 8-inch medical accessory kit is stowed in a
compartment on the spacecraft right side wall beside the lunar
module pilot couch. The medical kit contains three motion
sickness injectors, three pain suppression injectors, one two-
ounce bottle first aid ointment, two one-ounce bottles eye
drops, three nasal sprays, two compress bandages, 12 adhesive
bandages, one oral thermometer, and four spare crew biomedical
harnesses. Pills in the medical kit are 60 antibiotic, 12
nausea, 12 stimulant, 18 pain killer, 60 decongestant, 24
diarrhea, 72 aspirin and 40 antacid. Additionally, a small
medical kit containing four stimulant, eight diarrhea and
four pain killer pills, 12 aspirin, one bottle eye drops, two
compress bandages, eight decongestant pills, one automatic
injector containing a pain killer, one bottle nasal spray is
stowed in the lunar module flight data file compartment.

Survival Kit

The survival kit is stowed in two rucksacks in the right-hand forward equipment bay above the lunar module pilot.

Contents of rucksack No. 1 are: two combination survival lights, one desalter kit, three pair sunglasses, one radio beacon, one spare radio beacon battery and spacecraft connector cable, one knife in sheath, three water containers, and two containers of Sun lotion, two utility knives, three survival blankets and one utility netting.

Rucksack No. 2: one three-man life raft with CO_2 inflater, one sea anchor, two sea dye markers, three sun-bonnets, one mooring lanyard, three manlines and two attach brackets.

The survival kit is designed to provide a 48-hour post-landing (water or land) survival capability for three crewmen between 40° North and South latitudes.

Crew Food

More than 70 items comprise the food selection list of freeze-dried rehydratable, wet-pack and spoon-bowl foods. Balanced meals for five days have been packed in man/day wraps. Items similar to those in the daily menus have been packed in a snack pantry. The snack pantry permits the crew to locate easily a food item in a smorgasbord mode without having to "rob" a regular meal somewhere deep in a storage box.

Water for drinking and rehydrating food is obtained from two sources in the command module--a dispenser for drinking water and a water spigot at the food preparation station supplying water at about 155 or 55° F. The potable water dispenser squirts water continuously as long as the trigger is held down, and the food preparation spigot dispenses water in one-ounce increments.

A continuous-feed hand water dispenser similar to the one in the command module is used aboard the lunar module for cold-water rehydration of food packets stowed aboard the LM.

After water has been injected into a food bag, it is kneaded for about three minutes. The bag neck is then cut off and the food squeezed into the crewman's mouth. After a meal, germicide pills attached to the outside of the food bags are placed in the bags to prevent fermentation and gas formation. The bags are then rolled and stowed in waste disposal compartments.

NATIONAL AERONAUTICS AND SPACE ADMINISTRATION

WASHINGTON, D. C. 20546

BIOGRAPHICAL DATA

NAME: Alan B. Shepard, Jr. (Captain, USN), Apollo 14
prime crew Commander, NASA Astronaut

BIRTHPLACE AND DATE: Born Nov. 15, 1923, in East Derry, N.H.
His parents are Mr. and Mrs. Alan B. Shepard of East
Derry.

PHYSICAL DESCRIPTION: Brown hair; blue eyes; height: 5 feet
11 inches; weight: 165 pounds.

EDUCATION: Attended primary and secondary schools in East
Derry and Derry, N.H.; received a Bachelor of Science
degree from the United States Naval Academy in 1944
and an Honorary Master of Arts degree from Dartmouth
College in 1962.

MARITAL STATUS: Married to the former Louise Brewer of
Kennett Square, Pa. Her mother, Mrs. Julia Brewer,
resides in Wilmington, Del.

CHILDREN: Laura Shepard Snyder, July 2, 1947; Julie,
March 16, 1951.

ORGANIZATIONS: Fellow of the American Astronautical Society;
member of the Society of Experimental Test Pilots,
the Rotary, the Kiwanis, the Mayflower Society, the
Order of the Cincinnati, and the American Fighter
Aces; and honorary member of the Board of Directors
for the Houston School for Deaf Children.

SPECIAL HONORS: Awarded the Langley Medal (highest award of
the Smithsonian Institution) on May 5, 1964; the
Distinguished Flying Cross; the NASA Distinguished
Service Medal; the Navy Astronaut Wings; the Lambert
Trophy; the Kinchloe Trophy; the Cabot Award; and the
Collier Trophy.

-more-

EXPERIENCE: Shepard began his naval career, after graduation
          from Annapolis, on the destroyer COGSWELL deployed in
          the Pacific during World War II. He subsequently
          entered flight training at Corpus Christi, Tex., and
          Pensacola, Fla., and received his wings in 1947. His
          next assignment was with Fighter Squadron 42 at Norfolk,
          Va., and Jacksonville, Fla. He served several tours
          aboard aircraft carriers in the Mediterranean while
          with this squadron.

          In 1950, he attended the United States Navy Test Pilot
          School at Patuxent River, Md. After graduation, he
          participated in flight test work which included high-
          altitude tests to obtain data on light at different
          altitudes and on a variety of air masses over the
          American continent; and test and development experi-
          ments of the Navy's in-flight refueling system, carrier
          suitability trials of the F2H3 Banshee, and Navy trials
          of the first angled carrier deck. He was subsequently
          assigned to Fighter Squadron 193 at Moffett Field, Calif.,
          a night fighter unit flying Banshee jets. As operations
          officer of this squadron, he made two tours to the
          Western Pacific aboard the carrier ORISKANY.

          He returned to Patuxent for a second tour of duty and
          engaged in flight testing the F3H Demon, F8U Crusader,
          F4D Skyray, and F11F Tigercat. He was also project
          test pilot on the F5D Skylancer, and his last five
          months at Patuxent were spent as an instructor in the
          Test Pilot School.

          Upon completion of this tour of duty, he graduated
          from the Naval War College at Newport, R.I., and was
          subsequently assigned to the staff of the Commander-in-
          Chief, Atlantic Fleet, as aircraft readiness officer.

          He has logged more than 4,700 hours flying time--2,900
          hours in jet aircraft.

CURRENT ASSIGNMENT: Capt. Shepard was one of the Mercury
          astronauts named by NASA in April 1959, and he holds
          the distinction of being the first American to journey
          into space.

          On May 5, 1961, in the Freedom 7 spacecraft, he was
          launched by a Redstone vehicle on a ballistic tra-
          jectory suborbital flight--a flight which carried him
          to an altitude of 116 statute miles and to a landing
          point 302 statute miles down the Atlantic Missile
          Range.

He was designated Chief of the Astronaut Office in
1963 with responsibility for monitoring the coordi-
nation, scheduling, and control of all activities
involving NASA astronauts. This included monitoring
the development and implementation of effective
training programs to assure the flight readiness
of available pilot/non-pilot personnel for assignment
to crew positions on manned space flights; furnishing
pilot evaluations applicable to the design, con-
struction, and operations of spacecraft systems and
related equipment; and providing qualitative scientific
and engineering observations to facilitate overall
mission planning, formulation of feasible operational
procedures, and selection and conduct of specific
experiments for each flight.

In May 1969, he was restored to full flight status,
following correction of an inner ear disorder.
Capt. Shepard was subsequently named to serve as
spacecraft commander for the Apollo 14 flight.

####

NATIONAL AERONAUTICS AND SPACE ADMINISTRATION

WASHINGTON, D. C. 20546

BIOGRAPHICAL DATA

NAME: Stuart Allen Roosa (Major, USAF), Apollo 14 prime crew Command Module Pilot, NASA Astronaut

BIRTHPLACE AND DATE: Born Aug. 16, 1933, in Durango, Col. His parents, Mr. and Mrs. Dewey Roosa, now reside in Tucson.

PHYSICAL DESCRIPTION: Red hair; blue eyes; height: 5 feet 10 inches; weight: 155 pounds.

EDUCATION: Attended Justice Grade School and Claremore High School in Claremore, Okla.; studied at Oklahoma State University and the University of Arizona and was graduated with honors and a Bachelor of Science degree in Aeronautical Engineering from the University of Colorado.

MARITAL STATUS: His wife is the former Joan C. Barrett of Tupelo, Miss., and her mother, Mrs. John T. Barrett, resides in Sessums, Miss.

CHILDREN: Christopher A., June 29, 1959; John D., Jan. 2, 1961; Stuart A., Jr., March 12, 1962; Rosemary D., July 23, 1963.

RECREATIONAL INTERESTS: His hobbies are hunting, boating, and fishing.

EXPERIENCE: Roosa, a Major in the Air Force, has been on active duty since 1953. His last assignment was as an experimental test pilot at Edwards Air Force Base, Calif., from September 1965 to May 1966, subsequent to graduating from the Aerospace Research Pilots School in September 1965.

He was a maintenance flight test pilot at Olmsted Air Force Base, Pa., from July 1962 to August 1964, flying F-101 aircraft. He served as Chief of Service Engineering (AFLC) at Tachikawa Air Base, Japan for two years following graduation from the University of Colorado under the Air Force Institute of Technology Program. Prior to this tour of duty, he was assigned as a fighter pilot at Langley Air Force Base, Va., where he flew the F-84F and F-100 aircraft.

-more-

He attended Gunnery School at Del Rio and Luke Air
Force Bases and is a graduate of the Aviation Cadet
Program at Williams Air Force Base, Ariz., where he
received his flight training and commission in the
Air Force.

Since 1953 he has acquired 4,300 flying hours--3,900
in jet aircraft.

CURRENT ASSIGNMENT: Major Roosa is one of the 19 astronauts
    selected by NASA in April 1966. He was a member of the
    astronaut support crew for the Apollo 9 flight.

####

NATIONAL AERONAUTICS AND SPACE ADMINISTRATION

WASHINGTON, D. C. 20546

BIOGRAPHICAL DATA

NAME: Edgar Dean Mitchell (Commander, USN) Apollo 14 prime
crew Lunar Module Pilot, NASA Astronaut.

BIRTHPLACE AND DATE: Born in Hereford, Tex., on Sept. 17, 1930,
but considers Artesia, N.M., his hometown. His mother,
Mrs. Ernest Wagoner, now resides in Tahlequah, Okla.

PHYSICAL DESCRIPTION: Brown hair; green eyes; height: 5 feet
11 inches; weight: 180 pounds.

EDUCATION: Attended primary schools in Roswell, N.M., and is
a graduate of Artesia High School in Artesia, N.M.;
received a Bachelor of Science degree in Industrial
Management from the Carnegie Institute of Technology in
1952, a Bachelor of Science degree in Aeronautical Engineering from the U.S. Naval Postgraduate School in 1961, and
a Doctor of Science degree in Aeronautics/Astronautics
from the Massachusetts Institute of Technology in 1964.

MARITAL STATUS: Married to the former Louise Elizabeth Randall
of Muskegon, Mich., whose mother, Mrs. Winslow Randall,
now resides in Pittsburgh, Pa.

CHILDREN: Karlyn L., August 12, 1953; Elizabeth R., March
24, 1959.

RECREATIONAL INTERESTS: He enjoys handball and swimming, and
his hobbies are scuba diving and soaring.

ORGANIZATIONS: Member of the American Institute of Aeronautics
and Astronautics; the Society of Experimental Test Pilots;
Sigma Xi; and Sigma Gamma Tau.

EXPERIENCE: Cmdr. Mitchell's experience includes Navy operational flight, test flight, engineering, engineering
management, and some experience as a college instructor.
Mitchell came to the Manned Spacecraft Center after
graduating first in his class from the Air Force Aerospace
Research Pilot School where he was both student and part-
time instructor.

He entered the Navy in 1952 and completed his basic training at the San Diego Recruit Depot. In May 1953, after completing instruction at the Officers' Candidate School at Newport, R.I., he was commissioned as an Ensign. He completed his flight training in July 1954 at Hutchinson, Kans., and subsequently was assigned to Patrol Squadron 29 deployed to Okinawa.

From 1957 to 1958, he flew A3 aircraft while assigned to Heavy Attack Squadron Two deployed aboard the USS Bon Homme Richard and USS Ticonderoga; and he was a research project pilot with Air Development Squadron Five until 1959. His assignment from 1964 to 1965 was as Chief, Project Management Division of the Navy Field Office for Manned Orbiting Laboratory.

He has accumulated 3,700 hours flight time--1,600 hours in jets.

CURRENT ASSIGNMENT: Cmdr. Mitchell was in the group selected for astronaut training in April 1966. He served as a member of the astronaut support crew for Apollo 9 and as backup lunar module pilot for Apollo 10.

#####

-more-

NATIONAL AERONAUTICS AND SPACE ADMINISTRATION

WASHINGTON, D. C. 20546

BIOGRAPHICAL DATA

NAME: Eugene A. Cernan (Captain, USN) **Apollo 14 Backup Crew Commander, NASA Astronaut**

BIRTHPLACE AND DATE: Born in Chicago, Illinois, on March 14, 1932. His mother, Mrs. Andrew G. Cernan, resides in Bellwood, Illinois.

PHYSICAL DESCRIPTION: Brown hair; blue eyes; height: 6 feet; weight: 170 pounds.

EDUCATION: Graduated from Proviso Township High School in Maywood, Illinois; received a Bachelor of Science degree in Electrical Engineering from Purdue University and a Master of Science degree in Aeronautical Engineering from the U. S. Naval Postgraduate School; recipient of an Honorary Doctorate of Laws degree from Western State University College of Law in 1969 and an Honorary Doctorate of Engineering from Purdue University in 1970.

MARITAL STATUS: Married to the former Barbara J. Atchley of Houston, Texas.

CHILDREN: Teresa Dawn, March 4, 1963.

RECREATIONAL INTERESTS: His hobbies include gardening, and all sports activities.

ORGANIZATIONS: Member of the Society of Experimental Test Pilots; Tau Beta Pi, national engineering society; Sigma Xi, national science research society; and Phi Gamma Delta, national social fraternity.

SPECIAL HONORS: Awarded the NASA Distinguished Service Medal, the NASA Exceptional Service Medal, the Navy Distinguished Service Medal, the Navy Astronaut Wings, the Navy Distinguished Flying Cross, the National Academy of Television Arts and Sciences Special Trustees Award (1969), and an Honorary Lifetime Membership in the American Federation of Radio and Television Artists.

EXPERIENCE: Cernan, a United States Navy Captain, received his commission through the Navy ROTC program at Purdue. He entered flight training upon graduation.

He was assigned to Attack Squadrons 126 and 113 at the Miramar, California, Naval Air Station and subsequently attended the Naval Postgraduate School.

He has logged more than 3,800 hours flying time, with more than 3,600 hours in jet aircraft.

CURRENT ASSIGNMENT: Captain Cernan was one of the third group of astronauts selected by NASA in October 1963.

He occupied the pilot seat along side of command pilot Tom Stafford on the Gemini 9 mission. During this 3-day flight which began on June 3, 1966,

the spacecraft attained a circular orbit of 161 statute miles; the crew
used three different techniques to effect rendezvous with the previously
launched Augmented Target Docking Adapter; and Cernan logged two hours
and ten minutes outside the spacecraft in extravehicular activity. The
flight ended after 72 hours and 20 minutes with a perfect reentry and
recovery as Gemini 9 landed within 1½ miles of the prime recovery ship
USS WASP and 3/8 of a mile from the predetermined target point!

He subsequently served as backup pilot for Gemini 12 and as backup lunar
module pilot for Apollo VII.

Cernan was the lunar module pilot on Apollo X, May 18-26, 1969, the first
comprehensive lunar-orbital qualification and verification flight test of
an Apollo lunar module. He was accompanied on this 248,000 nautical mile
sojourn to the moon by Thomas P. Stafford (spacecraft commander) and
John W. Young (command module pilot). In accomplishing all of the assigned
objectives of this mission, Apollo X confirmed the operational performance,
stability, and reliability of the command-service module/lunar module
configuration during translunar coast, lunar orbit insertion, and lunar
module separation and descent to within 8 nautical miles of the lunar surface.
The latter maneuver involved employing all but the final minutes of the
technique prescribed for use in an actual lunar landing, and completing
critical evaluations of the lunar module propulsion systems and rendezvous
and landing radar devices in subsequent rendezvous and re-docking maneuvers.
In addition to demonstrating that man could navigate safely and accurately
in the moon's gravitational fields, Apollo X photographed and mapped
tentative landing sites for future missions.

This was Captain Cernan's second space flight giving him more than 264
hours and 24 minutes in space.

Captain   Cernan is currently assigned as backup spacecraft commander for
the forthcoming Apollo XIV flight.

#####

-more-

NATIONAL AERONAUTICS AND SPACE ADMINISTRATION

WASHINGTON, D. C. 20546

BIOGRAPHICAL DATA

NAME: Ronald E. Evans (Commander, USN) **Apollo 14 Backup Crew Command Module Pilot**, NASA Astronaut

BIRTHPLACE AND DATE: Born November 10, 1933, in St Francis, Kansas. His father, Mr. Clarence E. Evans, lives in St Francis and his mother, Mrs. Marie A. Evans, resides in Topeka, Kansas.

PHYSICAL DESCRIPTION: Brown hair; brown eyes; height: 5 feet $11\frac{1}{2}$ inches; weight: 160 pounds.

EDUCATION: Graduated from Highland Park High School in Topeka, Kansas; received a Bachelor of Science degree in Electrical Engineering from the University of Kansas in 1956 and a Master of Science degree in Aeronautical Engineering from the U. S. Naval Postgraduate School in 1964.

MARITAL STATUS: Married to the former Jan Pollom of Topeka, Kansas; her parents, Mr. and Mrs. Harry M. Pollom, reside in Salina, Kansas.

CHILDREN: Jaime D. (daughter), August 21, 1959; Jon P. (son), October 9, 1961.

RECREATIONAL INTERESTS: Hobbies include golfing, boating, swimming, fishing, and hunting.

ORGANIZATIONS: Member of Tau Beta Pi, Society of Sigma Xi, and Sigma Nu.

SPECIAL HONORS: Winner of eight Air Medals, the Viet Nam Service Medal, and the Navy Commendation Medal with combat distinguishing device.

EXPERIENCE: When notified of his selection to the astronaut program, Evans was on sea duty in the Pacific--assigned to VF-51 and flying F8 aircraft from the carrier USS TICONDEROGA during a period of seven months in Viet Nam combat operations.

He was a Combat Flight Instructor (F8 aircraft) with VF-124 from January 1961 to June 1962 and, prior to this assignment, participated in two WESTPAC aircraft carrier cruises while a pilot with VF-142. In June 1957, he completed flight training after receiving his commission as an Ensign through the Navy ROTC program at the University of Kansas.

Total flight time accrued during his military career is 3,500 hours-- 3,100 hours in jet aircraft.

CURRENT ASSIGNMENT: Commander Evans is one of the 19 astronauts selected by
    NASA in April 1966. He served as a member of the astronaut support crew
    for the Apollo VII and XI flights and has since been designated to serve
    as backup command module pilot for the forthcoming Apollo XIV mission.

#####

-more-

NATIONAL AERONAUTICS AND SPACE ADMINISTRATION

WASHINGTON, D. C. 20546

BIOGRAPHICAL DATA

NAME: Joe Henry Engle (Lieutenant Colonel, USAF) **Apollo 14 Backup Crew Lunar Module Pilot, NASA Astronaut**

BIRTHPLACE AND DATE: Born August 26, 1932, in Abilene, Kansas. His mother, Mrs. Abner E. Engle, resides in Chapman, Kansas.

PHYSICAL DESCRIPTION: Blond hair; hazel eyes; height: 5 feet $11\frac{1}{2}$ inches; weight: 155 pounds.

EDUCATION: Attended primary and secondary schools in Chapman, Kansas, and is a graduate of Dickinson County High School; received a Bachelor of Science degree in Aeronautical Engineering from the University of Kansas in 1955.

MARITAL STATUS: Married to the former Mary Catherine Lawrence of Mission Hills, Kansas; her parents are Mr. and Mrs. Ray E. Lawrence of Mission Hills.

CHILDREN: Laurie J., April 25, 1959; Jon L., May 9, 1962.

RECREATIONAL INTERESTS: His hobbies include hunting, athletics, and flying.

ORGANIZATIONS: Member of the Society of Experimental Test Pilots and the Theta Tau Fraternity.

SPECIAL HONORS: As a result of accomplishments as pilot of the NASA-USAF X-15 Research Rocket Plane, he received the Air Force Astronaut Wings and Distinguished Flying Cross and was named the Air Force Association's Outstanding Young Officer in 1964. That same year he was also selected by the U. S. Junior Chamber of Commerce as one of the Ten Outstanding Young Men in America. He received the American Institute of Aeronautics and Astronautics "Lawrence Sperry" Award in 1966 for experimental research in aerodynamics as test pilot of the X-15.

EXPERIENCE: Engle, an Air Force Lt Colonel, was an aerospace flight test pilot in the X-15 research program at Edwards Air Force Base, California, from June 1963 until his assignment to the Manned Spacecraft Center. Three of his sixteen flights in the X-15 exceeded an altitude of 50 miles (the designated altitude that qualifies a pilot for astronaut rating). Prior to that time, he was a test pilot in the Fighter Test Group at Edwards.

He received his commission in the Air Force through the AFROTC Program at the University of Kansas and entered flying school in 1957. Upon completion of flight training, he served with the 474th Fighter Day Squadron and the 309th Tactical Fighter Squadron at George Air Force Base, California. He has been stationed in Spain, Italy, and Denmark

during his military career and is a graduate of the USAF Experimental
Test Pilot School and the Air Force Aerospace Research Pilot School.

During his career he has logged more than 5,100 hours flight time--3,900
hours in jet aircraft.

CURRENT ASSIGNMENT: Lt. Col. Engle is one of the 19 astronauts selected by NASA in
April 1966. He is currently involved in training for future manned space
flights, and he was assigned as the lunar module pilot on spacecraft 2TV-1.
As a crewman on spacecraft 2TV-1, he participated in an eight-day thermal
vacuum test of the Apollo command module in June 1968. He also served
as a member of the astronaut support crew for the Apollo X flight.

Engle has since been designated to serve as backup lunar module pilot for
the forthcoming Apollo XIV mission.

#####

Flight Crew Health Stabilization Program

An expanded program for minimizing crew exposure to disease and illness was placed into effect 90 days prior to launch of the Apollo 14 crew. At 21 days prior to launch, closer medical surveillance of the crew and persons with whom they come in contact began and the number of persons having contact with the crew has been limited. Additionally, the crew has been limited to areas where microbial contamination is at a minimum.

Called the Flight Crew Health Stabilization Program, provisions are made in the program for crew epidemiology, immunology and clinical medicine.

Under the program, the prime and backup crews live in the crew quarters at the KSC Manned Spacecraft Operations Building during the final 21 days. The crew generally are limited to the primary areas of the MSOB and the Flight Crew Training Building, the Patrick AFB flight line and Pad 39A white room. Control and security provisions have been established for possible trips to Houston.

Special badging and security restrictions limit the crew personnel contacts to wives and those persons directly involved in training or mission preparations. Primary contacts - wives, backup crewmen, mission technicians and training personnel - have all had physical examinations and immunizations as part of the program.

APOLLO 14 FLAGS, LUNAR MODULE PLAQUE

The United States flag to be erected on the lunar surface measures 30 by 48 inches and will be deployed on a two-piece aluminum tube eight feet long. The folding horizontal bar which keeps the flag standing out from the staff on the airless Moon has been improved over the mechanisms used on Apollo 11 and 12.

The flag, made of nylon, will be stowed in the lunar module descent stage modularized equipment stowage assembly instead of in a thermal-protective tube on the LM front leg, as in Apollo 11 and 12.

Also carried on the mission and returned to Earth will be 25 United States flags, 50 individual state flags, flags of United States territories and flags of all United Nations members, each four by six inches.

A seven by nine-inch stainless steel plaque, similar to those flown on Apollos 11 and 12, will be fixed to the LM front leg. The plaque has on it the words "Apollo 14" with "Antares" beneath, January 1971, and the signatures of the three crewmen.

LUNAR RECEIVING LABORATORY (LRL)

The final phase of the back contamination program is completed in the MSC Lunar Receiving Laboratory. The crew and spacecraft are quarantined for a minimum of 21 days after completion of lunar EVA operations and are released based upon the completion of prescribed test requirements and results. The lunar sample will be quarantined for a period of 50 to 80 days depending upon results of extensive biological tests.

The LRL serves four basic purposes:

-    Quarantine of crew and spacecraft, the containment of lunar and lunar-exposed materials, and quarantine testing to search for adverse effects of lunar material upon terrestrial life.

-    The preservation and protection of the lunar samples.

-    The performance of time critical investigations.

-    The preliminary examination of returned samples to assist in an intelligent distribution of samples to principal investigators.

The LRL has the only vacuum system in the world which lets a man work through gloves leading directly into a chamber at pressures of about 10 billionth of an atmosphere. It has a low level radiation counting facility with a background count an order of magnitude better than other known counters. Additionally, it is a facility that can handle a large variety of biological specimens inside Class III biological cabinets designed to contain extremely hazardous pathogenic material.

The LRL covers 83,000 square feet of floor space and includes a Crew Reception Area (CRA), Vacuum Laboratory, Sample Laboratories (Physical and Bio-Science) and an Administrative and Support area. Special building systems are employed to maintain air flow into sample handling areas and the CRA, to sterilize liquid waste, and to incinerate contaminated air from the primary containment systems.

The biomedical laboratories provide for quarantine tests to determine the effect of lunar samples on terrestrial life. These tests are designed to provide data upon which to base the decision to release lunar material from quarantine.

Among the tests:

a.   Lunar material will be applied to 12 different
culture media and maintained under several environmental
conditions. The media will be observed for bacterial or
fungal growth. Detailed prelaunch inventories of the microbial
flora of the spacecraft and crew have been maintained so that any
living material found in the sample testing can be compared
against this list of potential contaminants taken to the Moon
by the crew or spacecraft.

b.   Six types of human and animal tissue culture cell
lines will be maintained in the laboratory and together with
embryonated eggs are exposed to the lunar material. Based on
cellular and/or other changes, the presence of viral material
can be established so that special tests can be conducted to
identify and isolate the type of virus present.

c.   Thirty-three species of plants and seedlings will
be exposed to lunar material. Seed germination, growth of
plant cells or the health of seedlings are then observed,
and histological, microbiological and biochemical techniques
are used to determine the cause of any suspected abnormality.

d.   A number of lower animals will be exposed to lunar
material, including germ-free mice, fish, birds, oysters,
shrimp, cockroaches, houseflies, planaria, paramecia and
euglena. If abnormalities are noted, further tests will be
conducted to determine if the condition is transmissible from
one group to another.

The crew reception area provides biological containment
for the flight crew and 12 support personnel. The nominal
occupancy is about 14 days but the facility is designed and
equipped to operate for considerably longer.

Sterilization and Release of the Spacecraft

Postflight testing and inspection of the spacecraft is
presently limited to investigation of anomalies which happened
during the flight. Generally, this entails some specific
testing of the spacecraft and removal of certain components of
systems for further analysis. The timing of postflight testing
is important so that corrective action may be taken for subsequent flights.

The schedule calls for the spacecraft to be returned
to port where a team will deactivate pyrotechnics, and flush
and drain fluid systems (except water). This operation will
be confined to the exterior of the spacecraft. The spacecraft
will then be flown to the LRL and placed in a special room for
storage, sterilization, and postflight checkout.

SATURN V LAUNCH VEHICLE

The Saturn V launch vehicle (SA-509) assigned to the Apollo 14 mission was developed under direction of the Marshall Space Flight Center, Huntsville, Ala. The vehicle is similar to those used in the missions of Apollo 8 through 13.

First Stage

The first stage (S-IC) of the Saturn V was built by the Boeing Co. at NASA's Michoud Assembly Facility, New Orleans, La. The stage's five F-1 engines develop about 7.6 million pounds of thrust at launch. Major stage components are the forward skirt, oxidizer tank, inter-tank structure, fuel tank, and thrust structure. Propellant to the five engines normally flows at a rate of 29,364.5 pounds (3,400 gallons) each second. One engine is rigidly mounted on the stage's centerline; the other four engines are mounted on a ring at $90°$ angles around the center engine. These outer engines are gimbaled to control the vehicle's attitude during flight.

Second Stage

The second stage (S-II) was built by the Space Division of the North American Rockwell Corp. at Seal Beach, Calif. Five J-2 engines develop a total of about 1.16 million pounds of thrust during flight. Major structural components are the forward skirt, liquid hydrogen and liquid oxygen tanks (separated by an insulated common bulkhead), a thrust structure and an interstage section that connects the first and second stages. The engines are mounted and used in the same arrangement as the first stage's F-1 engines: Four outer engines can be gimbaled; the center one is fixed.

Third Stage

The third stage (S-IVB) is built by the McDonnell Douglas Astronautics Co. at Huntington Beach, Calif. Major components are the aft interstage and skirt, thrust structure, two pro-pellant tanks with a common bulkhead, a forward skirt, and a single J-2 engine. The gimbaled engine has a maximum thrust of 230,000 pounds, and can be restarted in Earth orbit.

-more-

SPACECRAFT 82 FT.	**INSTRUMENT UNIT (IU)**
CM	
SM	Diameter: 21.7 feet
LM INSTRUMENT UNIT	Height: 3 feet
	Weight: 4,482 lbs.

INSTRUMENT UNIT (IU)

Diameter:	21.7 feet
Height:	3 feet
Weight:	4,482 lbs.

THIRD STAGE (S-IVB)

Diameter:	21.7 feet
Height:	59.3 feet
Weight:	260,080 lbs. fueled
	24,964 lbs. dry
Engine:	One J-2
Propellants:	Liquid Oxygen (191,532 lbs.; 20,228 gals.)
	Liquid Hydrogen (43,500 lbs.; 64,145 gals.)
Thrust:	199,500 lbs.
Interstage:	8,080 lbs.

SECOND STAGE (S-II)

Diameter:	33 feet
Height:	81.5 feet
Weight:	1,076,200 lbs. fueled
	78,050 lbs. dry
Engines:	Five J-2
Propellants:	Liquid Oxygen (836,120 lbs.; 88,215 gals.)
	Liquid Hydrogen (159,774 lbs.; 272,340 gals.)
Thrust:	924,200 to 1,165,000 lbs.
Interstages:	11,440 lbs.

FIRST STAGE (S-IC)

Diameter:	33 feet
Height:	138 feet
Weight:	4,949,100 lbs. fueled
	287,500 lbs. dry
Engines:	Five F-1
Propellants:	Liquid Oxygen (3,306,494 lbs.; 348,343 gals.)
	RP-1 Kerosene (1,435,647 lbs.; 215,330 gals.)
Thrust:	7,591,215 lbs. at liftoff

Diagram labels: SATURN V LAUNCH VEHICLE -281 FT. · SECOND STAGE (S-II) · THIRD STAGE (S-IVB) · FIRST STAGE (S-IC)

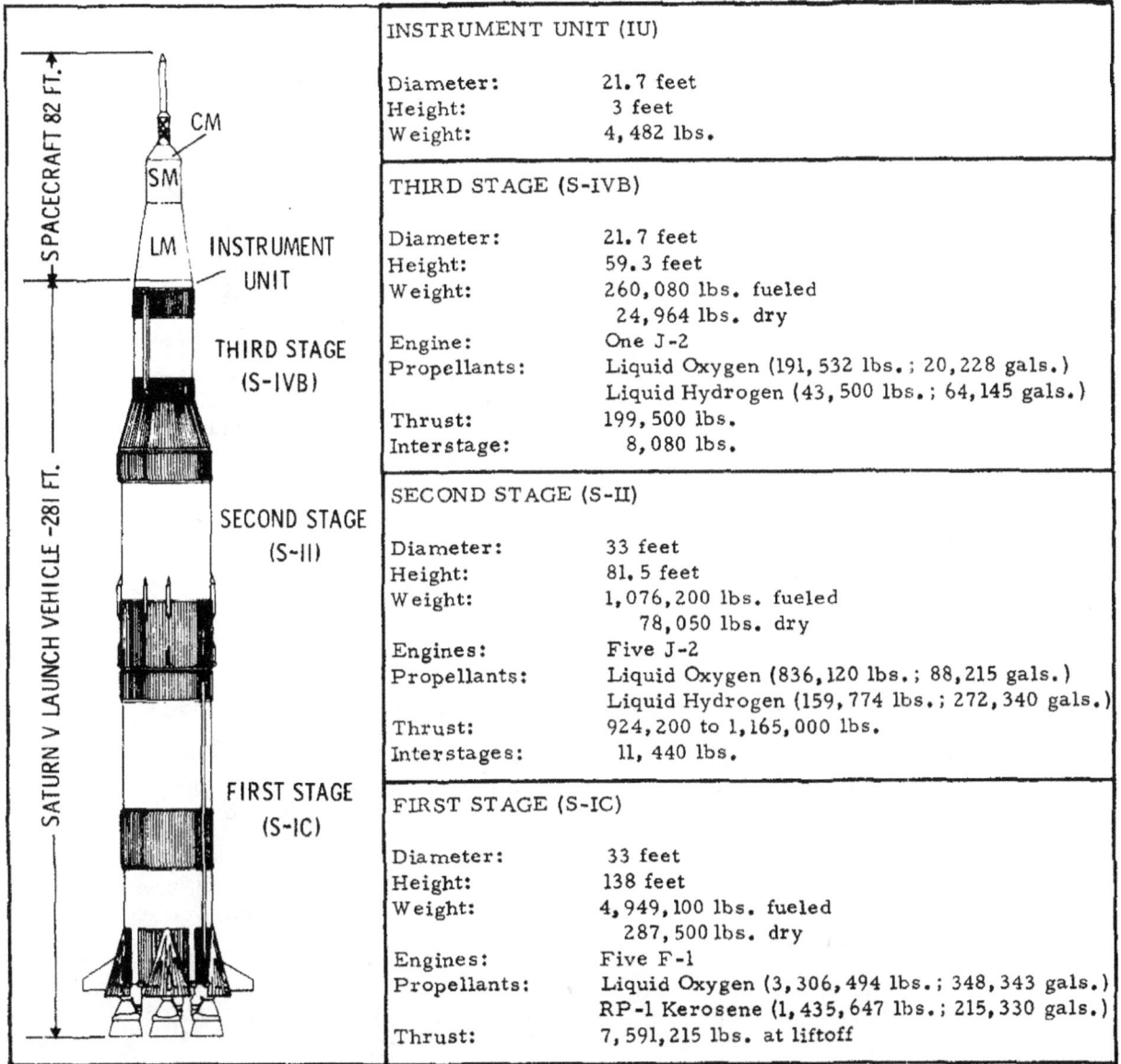

NOTE: Weights and measures given above are for the nominal vehicle configuration for Apollo. The figures may vary slightly due to changes before launch to meet changing conditions. Weights of dry stages and propellants do not equal total weight because frost and miscellaneous smaller items are not included in chart.

SATURN V LAUNCH VEHICLE

-more-

Instrument Unit

The instrument unit (IU), built by the International Business Machines Corp. at Huntsville, Ala., contains navigation, guidance and control equipment to steer the launch vehicle into its Earth orbit and into translunar trajectory. The six major systems are structural, thermal control, guidance and control, measuring and telemetry, radio frequency, and electric.

The instrument unit's inertial platform provides space-fixed reference coordinates and measures acceleration along three mutually perpendicular axes of a coordinate system. If the platform fails during boost, systems in the Apollo space-craft are programmed to provide guidance for the launch vehicle. After second stage ignition, the spacecraft commander could manually steer the vehicle if the launch vehicle's platform was lost.

Propulsion

The Saturn V has 37 propulsive units, with thrust ratings ranging from 70 pounds to more than 1.5 million pounds. The large main engines burn liquid propellants; the smaller units use solid or hypergolic propellants.

The five F-1 engines give the first stage a thrust range of from 7,591,215 pounds at liftoff to 8,904,285 pounds at center engine cutoff. Each F-1 engine weighs almost 10 tons, is more than 18 feet long, and has a nozzle exit diameter of nearly 14 feet. The engines each consume almost three tons of propellant every second.

The first stage also has eight solid-fuel retrorockets that fire to separate the first and second stages. Each retrorocket produces a thrust of 75,800 pounds for 0.54 seconds.

The second and third stage J-2 engine thrust varies from 184,841 to 232,263 pounds during flight. The 3,500-pound J-2 engine is more efficient than the F-1 engine because the J-2 burns high-energy liquid hydrogen. F-1 and J-2 engines are built by the Rocketdyne Division of the North American Rockwell Corp., Canoga Park, Calif.

The second stage also has four 23,000-pound-thrust solid fuel ullage rockets that settle liquid propellant in the bottom of the main tanks and help attain a "clean" separation from the first stage. Four retrorockets, located in the S-IVB's aft interstage, separate the S-II from the S-IVB. Two jettisonable ullage rockets settle propellant before engine ignition. Eight smaller engines in the two auxiliary propulsion system modules on the S-IVB stage provide three-axis attitude control.

Major Vehicle Changes

Three major changes have been made to the S-II stage for Apollo 14: A helium gas accumulator has been installed in the liquid oxygen (LOX) line of the center engine, a backup cutoff device for this engine, and a simplified 2-position propellant utilization valve on each of the 5 J-2 engines.

The accumulator will lower the frequency of the center engine LOX line to prevent unusually high oscillations (called the "pogo" effect) such as were recorded during Apollo 13, causing an early shutdown of the S-II's center engine, although the launch vehicle met all flight objectives.

The accumulator is a reservoir or cavity filled with helium gas that is a dampener for fluid pressure oscillations in the engine's LOX line. It lowers the resonant frequency of the center engine LOX line so that its vibrations will not couple with the vibrations of the thrust structure and engines. This prevents the three systems from oscillating in rhythm.

The backup cutoff device will shutdown the center engine in the unlikely event of pogo instability resulting from an accumulator failure and prevent excessive vibrations of the beam support of the center engine.

The J-2 valve controls the propellant mixture ratio to the engine, providing high thrust when needed during the first part of the burn when the stage is the heaviest and providing lower thrust during the end of the burn for more efficiency. The redesigned, pneumatically-actuated 2 position valve replaces a motor-driven, servo-controlled valve and depends only on an actuation command from the vehicle instrument unit rather than complicated stage electronics.

APOLLO SPACECRAFT

The Apollo spacecraft for the Apollo 14 mission consists of the command module, service, module, lunar module, a spacecraft-lunar module adapter (SLA) and a launch escape system. The SLA houses the lunar module and serves as a mating structure between the Saturn V instrument unit and the SM.

Launch Escape System (LES) -- The function of the LES is to propel command module to safety in an aborted launch. It has three solid-propellant rocket motors: a 147,000-pound-thrust launch escape system-motor, a 2,400-pound-thrust pitch control motor, and a 31,500-pound-thrust tower jettison motor. Two canard vanes deploy to turn the command module aerodynamically to an attitude with the heat-shield forward. The system is 33 feet tall and four feet in diameter at the base, and weighs 9,025 pounds.

Command Module (CM) -- The command module is a pressure vessel encased in heat shields, cone-shaped, weighing 12,694 pounds at launch.

The command module consists of a forward compartment which contains two reaction control engines and components of the Earth landing system; the crew compartment or inner pressure vessel containing crew accomodations, controls and displays, and many of the spacecraft systems; and the aft compartment housing ten reaction control engines, propellant tankage, helium tanks, water tanks, and the CSM umbilical cable. The crew compartment contains 210 cubic feet of habitable volume.

Heat-shields around the three compartments are made of brazed stainless steel honeycomb with an outer layer of phenolic epoxy resin as an ablative material.

The CSM and LM are equipped with the probe-and-drogue docking hardware. The probe assembly is a powered folding coupling and impact attentuating device mounted in the CM tunnel that mates with a conical drogue mounted in the LM docking tunnel. After the 12 automatic docking latches are checked following a docking maneuver, both the probe and drogue are removed to allow crew transfer between the CSM and LM.

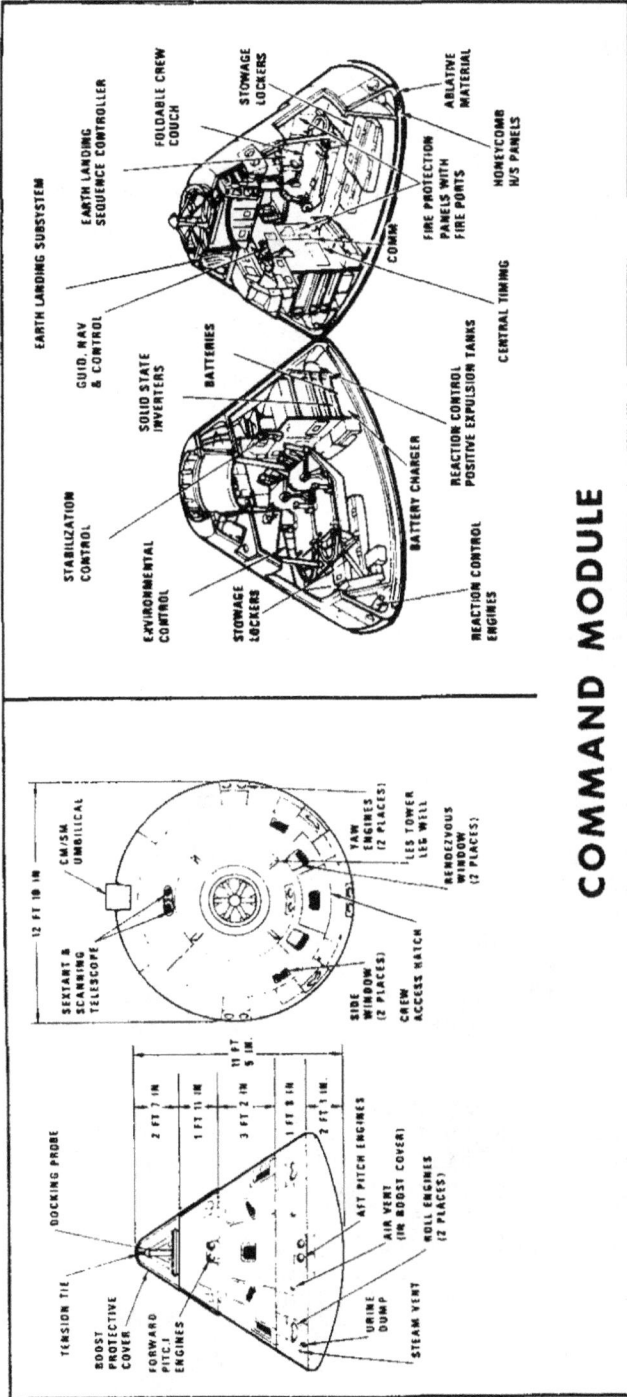

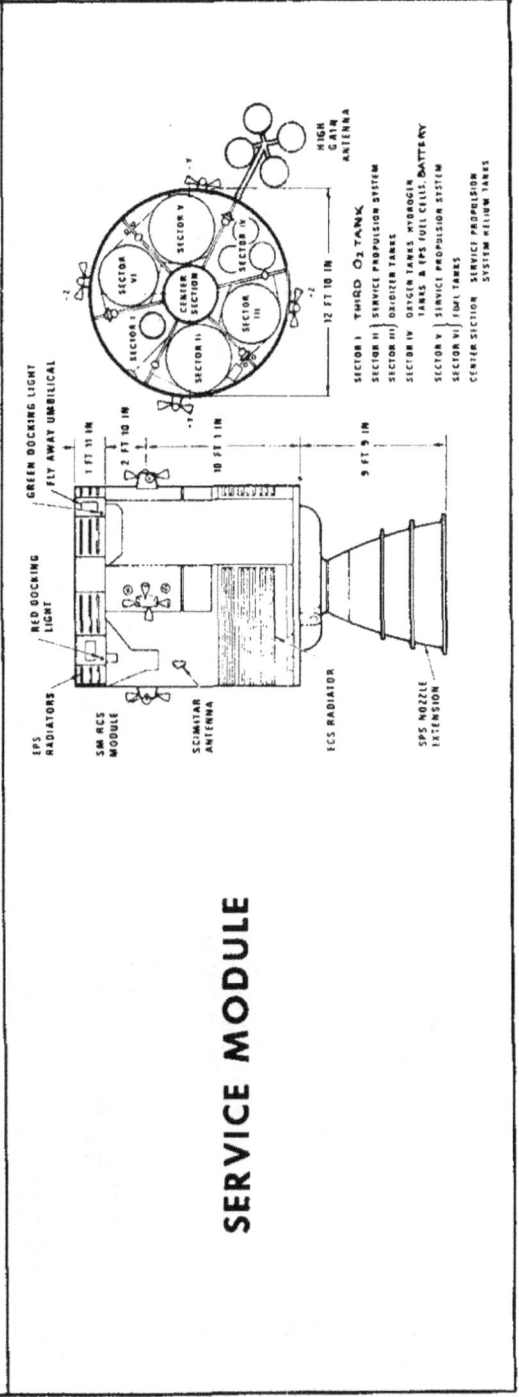

COMMAND MODULE

SERVICE MODULE

    Service Module (SM) -- At launch, the service module
for the Apollo 14 mission will weigh 51,744 pounds. Aluminum
honeycomb panels one-inch thick form the outer skin, and
milled aluminum radial beams separate the interior into six
sections around a central cylinder containing two helium
spheres, four sections containing service propulsion system
fuel-oxidizer tankage, another containing fuel cells,
cryogenic oxygen and hydrogen, and one sector essentially
empty.

    Spacecraft-LM Adapter (SLA) Structure -- The spacecraft
LM adapter is a truncated cone 28 feet long tapering from
260 inches diameter at the base to 154 inches at the forward
end at the service module mating line. The SLA weighs 4,061
pounds and houses the LM during launch and Earth orbital
flight.

Command-Service Module Modifications

    Following the aborted Apollo 13 lunar landing mission in
April 1970, the Apollo 13 Review Board recommended changes to
the command and service modules aimed at enhancing the spacecraft's ability to return a crew safely to Earth in case of
an emergency.

    The Apollo 13 abort was caused by a short circuit and
wiring overheating in one of the service module cryogenic
oxygen tanks. This caused a tank rupture and loss of the
prime oxygen supply for cabin pressurization, breathing and
oxygen reactant for the fuel cells. The incident occurred
about two-thirds of the way to the Moon. Because the service
propulsion system could not be assumed to be undamaged, the
LM propulsion was used for the balance of the mission. The
lunar module also was used as a "lifeboat" for its oxygen,
battery, and water supplies.

    The major changes to the command/service modules include
adding a third cryogenic oxygen tank installed in a heretofore
empty bay of the service module, addition of an auxilliary
battery in the service module as a backup in case of fuel cell
failure, removal of destratification fans in the cryogenic
oxygen tanks and removal of thermostat switches from the
oxygen tank heater circuits. Provision for stowage of an
emergency five-gallon supply of drinking water has been
added to the command module.

The third oxygen tank is in service module sector 1 on
the opposite side of the spacecraft from the other two tanks.
An isolation valve allows the third tank to be isolated from
the fuel cells and from the other two tanks in an emergency,
and to feed only the command module environmental control
system.

Internal oxygen tank wiring is now enclosed in stainless
steel conduit instead of the Teflon insulation previously
used, and a third heater element has been added to each tank.
These heaters may be turned on one, two or three at a time
instead of two heaters on or off as in earlier tanks. A
sensor has been provided to read the heater assembly temperature and the bulk temperature sensor has been relocated to
improve its accuracy. Onboard and telemetered readouts for
both sensors are provided.

The oxygen tank quantity gauge probe was changed from
aluminum to stainless steel, and all previously soldered
joints have been replaced with brazed joints.

The auxiliary battery in service module sector 4 is a
400-ampere-hour silver oxide/zinc nonrechargeable type weighing
135 pounds. The battery is identical to the four lunar module
descent stage batteries.

The emergency command module water supply consists of
five one-gallon plastic bags wrapped in beta cloth and
packaged in a stowage bag together with fill hose, valves,
and drinking nozzle. The stowage bag is kept in a locker
on the command module aft bulkhead. In some emergencies,
powering down causes low temperatures and freezing of the
water in the CM storage tank. Before freezing occurs, the
water bags would be filled. Prior to entry, any water in
the emergency water supply will be dumped overboard through
the waste management system.

(Additional detailed information on command module and
lunar module systems and subsystems is available in reference
documents at query desks at KSC and MSC News Centers.)

Lunar Module (LM)

The lunar module is a two-stage vehicle designed for space operations near and on the Moon. The lunar module stands 22 feet 11 inches high and is 31 feet wide (diagonally across landing gear.) The ascent and descent stages of the LM operate as a unit until staging, when the ascent stage functions as a single spacecraft for rendezvous and docking with the CM.

Ascent Stage --Three main sections make up the ascent stage: the crew compartment, midsection, and aft equipment bay. Only the crew compartment and midsection are pressurized (4.8 psig). The cabin volume is 235 cubic feet (6.7 cubic meters). The stage measures 12 feet four inches high by 14 feet one inch in diameter. The ascent stage has six substructural areas: crew compartment, midsection, aft equipment bay, thrust chamber assembly cluster supports, antenna supports, and thermal and micrometeoroid shield.

The cylindrical crew compartment is 92 inches (2.35 m) in diameter and 42 inches (1.07 m) deep. Two flight stations are equipped with control and display panels, armrests, body restraints, landing aids, two front windows, an overhead docking window, and an alignment optical telescope in the center between the two flight stations. The habitable volume is 160 cubic feet.

A tunnel ring atop the ascent stage meshes with the command module docking latch assemblies. During docking, the CM docking ring and latches are aligned by the LM drogue and the CSM probe.

The docking tunnel extends downward into the midsection 16 inches (40 cm). The tunnel is 32 inches (81 cm) in diameter and is used for crew transfer between the CSM and LM. The upper hatch on the inboard end of the docking tunnel opens inward and cannot be opened without equalizing pressure on both hatch surfaces.

A thermal and micrometeoroid shield of multiple layers of Mylar and a single thickness of thin aluminum skin encases the entire ascent stage structure.

As a result of the Apollo 13 mission, wiring provisions were made to the LM ascent stage to permit power transfer from LM to CM after LM descent stage separation.

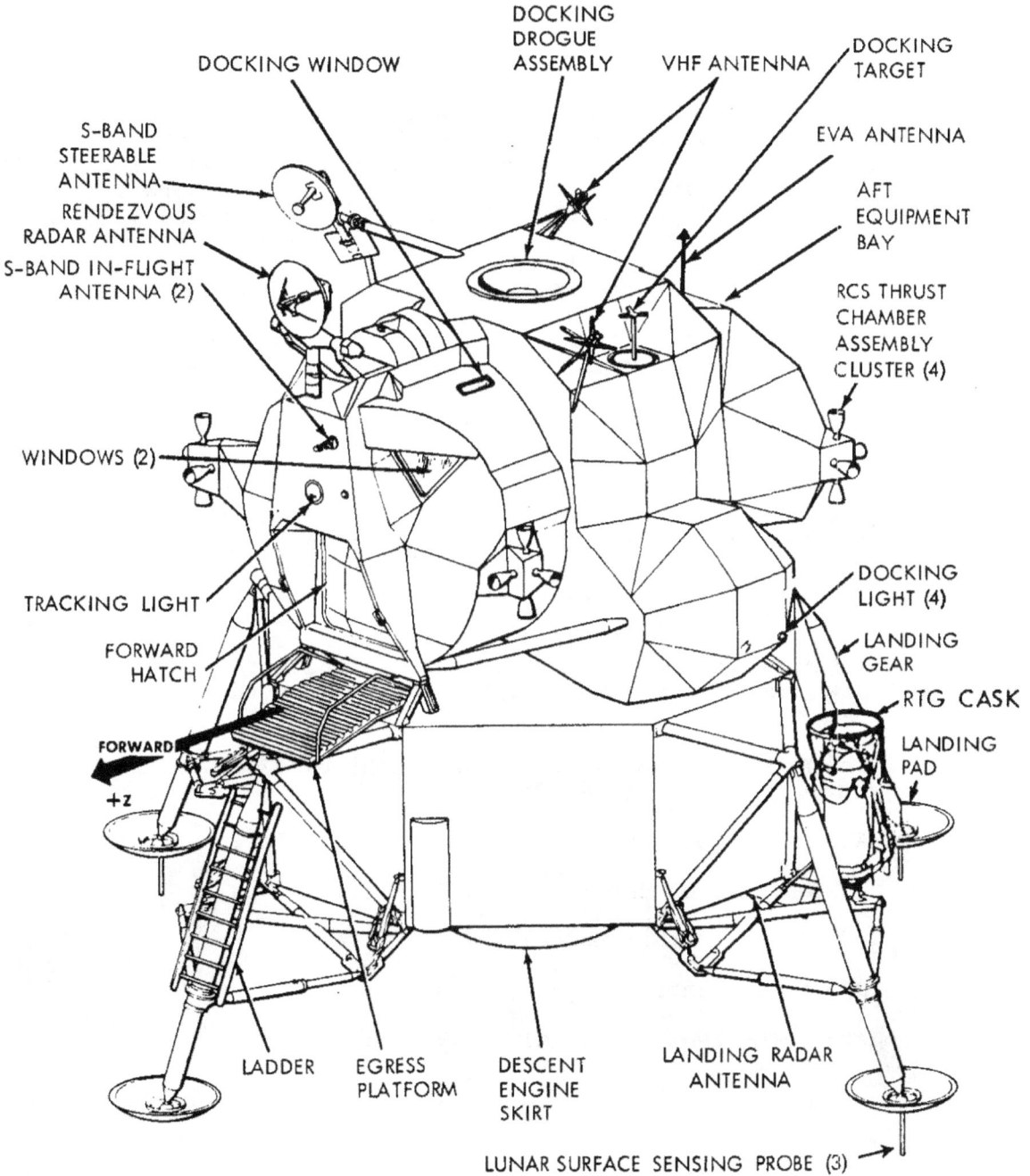

DOCKING WINDOW

DOCKING DROGUE ASSEMBLY

VHF ANTENNA

DOCKING TARGET

S-BAND STEERABLE ANTENNA

RENDEZVOUS RADAR ANTENNA

S-BAND IN-FLIGHT ANTENNA (2)

EVA ANTENNA

AFT EQUIPMENT BAY

RCS THRUST CHAMBER ASSEMBLY CLUSTER (4)

WINDOWS (2)

TRACKING LIGHT

FORWARD HATCH

FORWARD

+Z

DOCKING LIGHT (4)

LANDING GEAR

RTG CASK

LANDING PAD

LADDER

EGRESS PLATFORM

DESCENT ENGINE SKIRT

LANDING RADAR ANTENNA

LUNAR SURFACE SENSING PROBE (3)

LUNAR MODULE

-more-

Descent Stage --The descent stage center compartment
houses the descent engine, and descent propellant tanks
are housed in the four bays around the engine. Quadrant II
contains ALSEP. The radioisotope thermoelectric generator
(RTG) is externally mounted. Quadrant IV contains the
MESA. The descent stage measures ten feet seven inches
high by 14 feet one inch in diameter and is encased in the
Mylar and aluminum alloy thermal and micrometeoroid shield.

The LM egress platform or "porch" is mounted on the
forward outrigger just below the forward hatch. A ladder
extends down the forward landing gear strut from the porch
for crew lunar surface operations.

The landing gear struts are released explosively and
are extended by springs. They provide lunar surface landing
impact attenuation. The main struts are filled with crush-
able aluminum honeycomb for absorbing compression loads.
Footpads 37 inches (0.95 m) in diameter at the end of each
landing gear provide vehicle support on the lunar surface.

Each pad (except forward pad) is fitted with a 68-inch-
long lunar surface sensing probe which signals the crew to
shut down the descent engine upon contact with the lunar
surface.

The Apollo 14 LM has a launch weight of 33,680 pounds.
The weight breakdown is as follows:

Ascent stage, dry	4,691 lbs.	Includes water and oxygen; no crew
Descent stage, dry	4,716 lbs.	
RCS propellants (loaded)	633 lbs.	
DPS propellants (loaded)	18,415 lbs.	
APS propellants (loaded)	5,225 lbs.	
	33,680 lbs.	

MANNED SPACE FLIGHT NETWORK

The worldwide Manned Space Flight Network (MSFN) provides nearly continuous communications with the astronauts, launch vehicle, and spacecraft from liftoff to splashdown. After the flight, the network will continue the link between Earth and the Apollo experiments on the lunar surface.

The MSFN is maintained and operated by the NASA Goddard Space Flight Center, Greenbelt, Md., under the direction of NASA's Office of Tracking and Data Acquisition. If the Houston Mission Control Center is seriously impaired for an extended time, Goddard becomes the emergency control center.

The MSFN employs 11 ground tracking stations equipped with 30- and 85-foot antennas, an instrumented tracking ship, Vanguard, and four instrumented aircraft. For Apollo 14, the network will be augmented by the 210-foot antenna system at Goldstone, Calif.

NASA Communications Network (NASCOM). The tracking network is linked together by the NASA Communications Network. All information flows to and from MCC Houston and the Apollo spacecraft over this communications system.

The NASCOM consists of more than two million circuit miles, using satellites, submarine cables, land lines, microwave systems, and high frequency radio facilities. NASCOM control center is located at Goddard. Regional communication switching centers are in London; Madrid; Canberra, Australia; Honolulu; and Guam.

Three Intelsat communications satellites will be used for Apollo 14. One satellite over the Atlantic will link Goddard with Ascension Island and the Vanguard tracking ship. Another Atlantic satellite will provide a direct link between Madrid and Goddard for TV signals received from the spacecraft. The third satellite over the mid-Pacific will link Carnarvon, Canberra, and Hawaii with Goddard through a ground station at Brewster Flats, Wash.

Mission Operations. The Merritt Island tracking station supports prelaunch tests, the terminal countdown, as well as for flights.

-more-

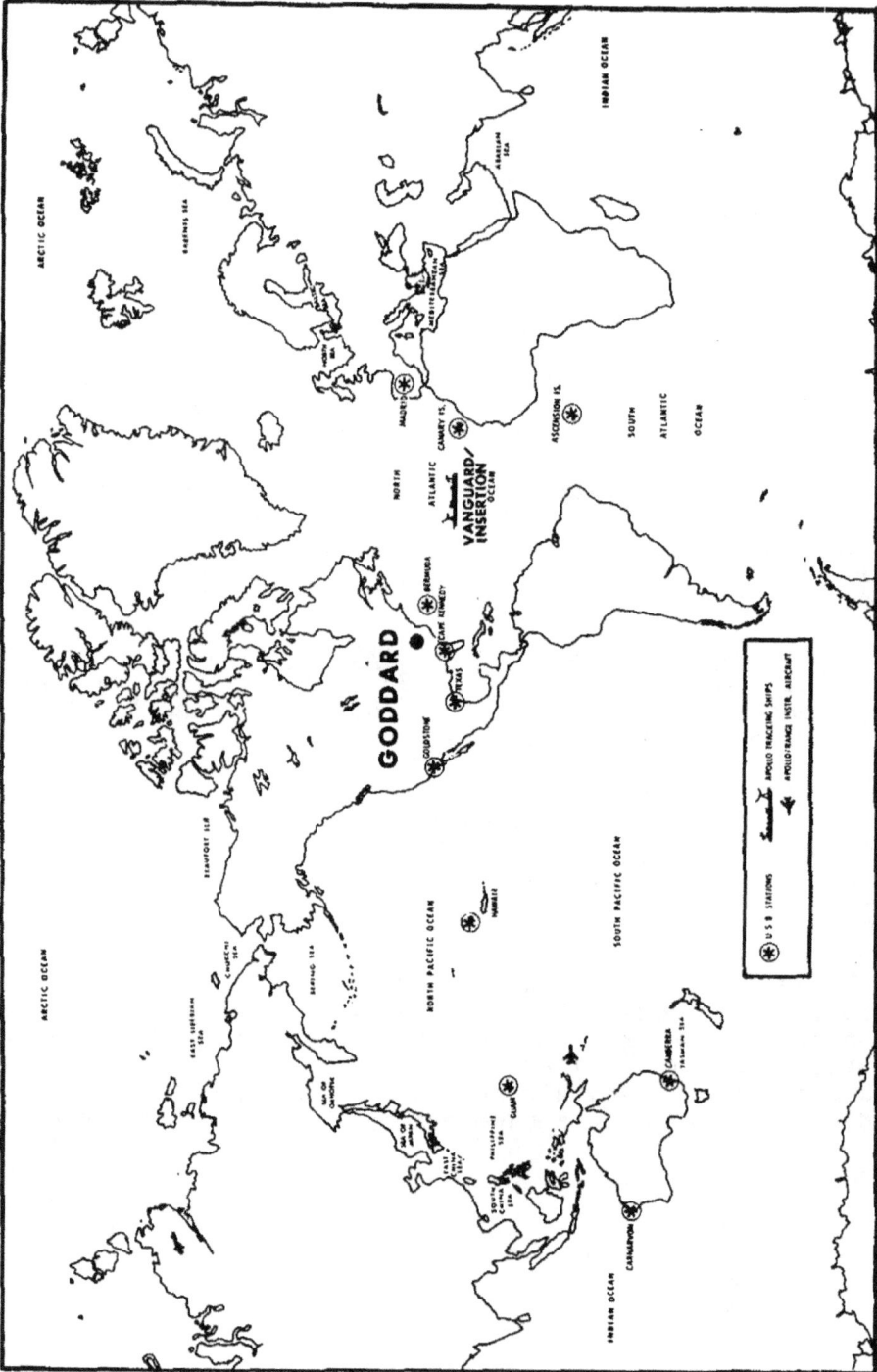

MANNED SPACE FLIGHT TRACKING NETWORK

The USNS Vanguard will perform tracking, telemetry, and communication functions for the launch phase and Earth orbit insertion. Vanguard will be stationed about 1,000 miles south-east of Bermuda.

During the Apollo 14 TLI maneuver, two Apollo Range Instrumentation Aircraft (ARIA) will record telemetry data from Apollo and relay voice communication between the astronauts and the Mission Control Center at Houston. The ARIA will be located between Australia and Hawaii.

Approximately one hour after the spacecraft has been injected into its translunar trajectory (some 10,000 miles from the Earth), three prime tracking stations spaced nearly equidistant around the Earth will take over tracking and communicating with Apollo.

Each of the prime stations, located at Goldstone, Madrid, and Canberra, has a dual system for use when tracking the command module in lunar orbit and the lunar module in separate flight paths or at rest on the Moon. These stations are equipped with 85-foot antennas.

For reentry, two ARIA will be deployed to the landing area to relay communications between Apollo and Mission Control at Houston and provide position information on the spacecraft after the blackout phase of reentry has passed.

Television Transmissions. Through the journey to the Moon and return, television will be received from the space-craft at the three prime stations. In addition, the 210-foot antenna at Goldstone (a unit of NASA's Deep Space Network) will augment the television coverage while Apollo 14 is near and on the Moon.

For color TV, the signal must be converted at the MSC Houston. A black and white version of the color signal can be released locally from the stations in Spain and Australia.

PROGRAM MANAGEMENT

The Apollo Program is the responsibility of the Office of Manned Space Flight (OMSF), National Aeronautics and Space Administration, Washington, D.C. Dale D. Myers is Associate Administrator for Manned Space Flight.

NASA Manned Spacecraft Center (MSC), Houston, is responsible for development of the Apollo spacecraft, flight crew training, and flight control. Dr. Robert R. Gilruth is Center Director.

NASA Marshall Space Flight Center (MSFC), Huntsville, Ala., is responsible for development of the Saturn launch vehicles. Dr. Eberhard F. M. Rees is Center Director.

NASA John F. Kennedy Space Center (KSC), Fla., is responsible for Apollo/Saturn launch operations. Dr. Kurt H. Debus is Center Director.

The NASA Office of Tracking and Data Acquisition (OTDA) directs the program of tracking and data flow on Apollo. Gerald M. Truszynski is Associate Administrator for Tracking and Data Acquisition.

NASA Goddard Space Flight Center (GSFC), Greenbelt, Md., manages the Manned Space Flight Network and Communications Network. Dr. John F. Clark is Center Director.

The Department of Defense is supporting NASA in Apollo 14 during launch, tracking and recovery operations. The Air Force Eastern Test Range is responsible for range activities during launch and down-range tracking. Recovery operations include the use of recovery ships and Navy and Air Force aircraft.

Apollo/Saturn Officials

NASA Headquarters

Dr. Rocco A. Petrone	Apollo Program Director, OMSF
Chester M. Lee,(Capt., USN, Ret.)	Apollo Mission Director, OMSF
Robert B. Sheridan	Apollo 14 Mission Engineer, OMSF
Dr. James W. Humphreys, Jr., (Maj. Gen., USAF MC, Ret.)	Director of Life Sciences, OMSF
John K. Holcomb, (Capt., USN, Ret.)	Director of Apollo Operations, OMSF
Lee R. Scherer,(Capt., USN, Ret.)	Director of Apollo Lunar Exploration, OMSF
James C. Bavely	Chief of Network Operations Branch, OTDA

Marshall Space Flight Center

Erich W. Neubert	Deputy Center Director, Technical (Acting)
R. W. Cook	Deputy Center Director, Management
Lee B. James	Director, Program Management
Dr. F. A. Speer	Manager, Mission Operations Office
Richard G. Smith	Manager, Saturn Program Office
Matthew W. Urlaub	Manager, S-IC Stage, Saturn Program Office
William F. LaHatte	Manager, S-II Stage, Saturn Program Office
Charles H. Meyers	Manager (Acting), S-IVB Stage, Saturn Program Office
Frederich Duerr	Manager, Instrument Unit, Saturn Program Office
William D. Brown	Manager, Engine Program Office

Kennedy Space Center

Miles J. Ross	Deputy Center Director
Walter J. Kapryan	Director of Launch Operations
Raymond L. Clark	Director of Technical Support
Brig. Gen. Thomas W. Morgan,(USAF)	Apollo/Skylab Program Manager
Dr. Robert H. Gray	Deputy Director, Launch Operations
Dr. Hans F. Gruene	Director, Launch Vehicle Operations
John J. Williams	Director, Spacecraft Operations
Paul C. Donnelly	Launch Operations Manager
Isom A. Rigell	Deputy Director for Engineering

Manned Spacecraft Center

Dr. Christopher C. Kraft, Jr.	Deputy Center Director
Col. James A. McDivitt, (USAF)	Manager, Apollo Spacecraft Program
Donald K. Slayton	Director, Flight Crew Operations
Sigurd A. Sjoberg	Director, Flight Operations
M. P. Frank	Flight Director
Milton L. Windler	Flight Director
Gerald Griffin	Flight Director
Eugene F. Kranz	Flight Director
Dr. Charles A. Berry	Director, Medical Research and Operations

Goddard Space Flight Center

Ozro M. Covington — Director of Manned Flight Support

William P. Varson — Chief, Manned Flight Planning & Analysis Division

H. William Wood — Chief, Manned Flight Operations Division

Tecwyn Roberts — Chief, Manned Flight Engineering Division

L. R. Stelter — Chief, NASA Communications Division

Department of Defense

Maj. Gen. David M. Jones, (USAF) — DOD Manager of Manned Space Flight Support Operations, Commander of USAF Eastern Test Range

Rear Adm. Wm. S. Guest, (USN) — Deputy DOD Manager of Manned Space Flight Support Operations, Commander Task Force 140, Atlantic Recovery Area

Rear Adm. Donald C. Davis, (USN) — Commander Task Force 130, Pacific Recovery Area

Col. Kenneth J. Mask, (USAF) — Director of DOD Manned Space Flight Support Office

Brig. Gen. Frank K. Everest, (USAF) — Commander Aerospace Rescue and Recovery Service

CONVERSION TABLE

	Multiply	By	To Obtain
Distance:	feet	0.3048	meters
	meters	3.281	feet
	kilometers	3281	feet
	kilometers	0.6214	statute miles
	statute miles	1.609	kilometers
	nautical miles	1.852	kilometers
	nautical miles	1.1508	statute miles
	statute miles	0.86898	nautical miles
	statute miles	1760	yards
Velocity:	feet/sec	0.3048	meters/sec
	meters/sec	3.281	feet/sec
	meters/sec	2.237	statute mph
	feet/sec	0.6818	statute miles/hr
	feet/sec	0.5925	nautical miles/hr
	statute miles/hr	1.609	km/hr
	nautical miles/hr (knots)	1.852	km/hr
	km/hr	0.6214	statute miles/hr
Liquid measure, weight:	gallons	3.785	liters
	liters	0.2642	gallons
	pounds	0.4536	kilograms
	kilograms	2.205	pounds
Volume:	cubic feet	0.02832	cubic meters
Pressure:	pounds/sq. inch	70.31	grams/sq. cm

-end-

NASA-KSC JAN/71

www.ingramcontent.com/pod-product-compliance
Lightning Source LLC
Chambersburg PA
CBHW051224200326
41519CB00025B/7237